Undergraduate Texts in Mathematics

Springer

New York
Berlin
Heidelberg
Hong Kong
London
Milan
Paris
Tokyo

Undergraduate Texts in Mathematics

Abbott: Understanding Analysis.
Anglin: Mathematics: A Concise History and Philosophy.
Readings in Mathematics.
Anglin/Lambek: The Heritage of Thales.
Readings in Mathematics.
Apostol: Introduction to Analytic Number Theory. Second edition.
Armstrong: Basic Topology.
Armstrong: Groups and Symmetry.
Axler: Linear Algebra Done Right. Second edition.
Beardon: Limits: A New Approach to Real Analysis.
Bak/Newman: Complex Analysis. Second edition.
Banchoff/Wermer: Linear Algebra Through Geometry. Second edition.
Berberian: A First Course in Real Analysis.
Bix: Conics and Cubics: A Concrete Introduction to Algebraic Curves.
Brémaud: An Introduction to Probabilistic Modeling.
Bressoud: Factorization and Primality Testing.
Bressoud: Second Year Calculus.
Readings in Mathematics.
Brickman: Mathematical Introduction to Linear Programming and Game Theory.
Browder: Mathematical Analysis: An Introduction.
Buchmann: Introduction to Cryptography.
Buskes/van Rooij: Topological Spaces: From Distance to Neighborhood.
Callahan: The Geometry of Spacetime: An Introduction to Special and General Relavitity.
Carter/van Brunt: The Lebesgue–Stieltjes Integral: A Practical Introduction.
Cederberg: A Course in Modern Geometries. Second edition.
Childs: A Concrete Introduction to Higher Algebra. Second edition.
Chung/AitSahlia: Elementary Probability Theory: With Stochastic Processes and an Introduction to Mathematical Finance. Fourth edition.
Cox/Little/O'Shea: Ideals, Varieties, and Algorithms. Second edition.
Croom: Basic Concepts of Algebraic Topology.
Curtis: Linear Algebra: An Introductory Approach. Fourth edition.
Daepp/Gorkin: Reading, Writing, and Proving: A Closer Look at Mathematics.
Devlin: The Joy of Sets: Fundamentals of Contemporary Set Theory. Second edition.
Dixmier: General Topology.
Driver: Why Math?
Ebbinghaus/Flum/Thomas: Mathematical Logic. Second edition.
Edgar: Measure, Topology, and Fractal Geometry.
Elaydi: An Introduction to Difference Equations. Second edition.
Erdős/Surányi: Topics in the Theory of Numbers.
Estep: Practical Analysis in One Variable.
Exner: An Accompaniment to Higher Mathematics.
Exner: Inside Calculus.
Fine/Rosenberger: The Fundamental Theory of Algebra.
Fischer: Intermediate Real Analysis.
Flanigan/Kazdan: Calculus Two: Linear and Nonlinear Functions. Second edition.
Fleming: Functions of Several Variables. Second edition.
Foulds: Combinatorial Optimization for Undergraduates.
Foulds: Optimization Techniques: An Introduction.

(continued after index)

J. David Logan

Applied Partial Differential Equations

Second Edition

With 40 Illustrations

J. David Logan
Department of Mathematics
University of Nebraska at Lincoln
Lincoln, NE 68588-0323
USA
dlogan@math.unl.edu

Mathematics Subject Classification (2000): 35-01

Library of Congress Cataloging-in-Publication Data
Logan, J. David (John David)
Applied partial differential equations / J. David Logan—2nd ed.
p. cm.—(Undergraduate texts in mathematics)
Includes bibliographical references and index.
ISBN 0-387-20935-2 (alk. paper). — ISBN 0-387-20953-0 (pbk. : alk. paper)
1. Differential equations, Partial. I. Title. II. Series.
QA377.L578 2004
515'.353—dc22 2004041692

ISBN 0-387-20935-2 (hardcover) Printed on acid-free paper.
ISBN 0-387-20953-0 (softcover)

Printed in the United States of America. (BPR/EBP)

9 8 7 6 5 4 3 2 1 SPIN 10975848 (hardcover) SPIN 10952818 (softcover)

Springer-Verlag is part of *Springer Science + Business Media*

springeronline.com

Preface to the Second Edition

This textbook is for the standard, one-semester, junior–senior course that often goes by the title "Elementary Partial Differential Equations" or "Boundary Value Problems." The audience consists of students in mathematics, engineering, and the sciences. This text gives derivations of the standard equations of engineering and science (including the advection equation, the diffusion equation, the wave equation, and the Laplace's equation) and methods for solving those equations on bounded and unbounded domains. Methods include eigenfunction expansions (separation of variables), integral transforms (Fourier and Laplace), characteristic methods, and finite difference methods. There is strong emphasis on modeling and applications throughout. The prerequisites are calculus and post-calculus differential equations.

There are several excellent existing texts, but most of them are long. This text was written to provide a brief, one-semester introduction to partial differential equations. It is limited in both scope and depth compared with existing books, yet it covers the main topics. The frontiers of mathematics and science are expanding rapidly, and a one-semester course must try to advance students to a level where they can reach these boundaries more quickly than in the past. Not every traditional topic can be examined in great detail. An example is the method of separation of variables, which plays a dominant role in most texts; a few well-chosen illustrations of the method should suffice.

The level of exposition in this text is slightly higher than one usually encounters in the post-calculus differential equations course. The philosophy is that a student should progress in the ability to read mathematics. Elementary calculus and ordinary differential equations texts contain many examples and detailed calculations, but advanced mathematics and science books leave a lot to the reader. This text leaves some of the easy details to supply to the reader. The student is encouraged as part of the learning process to fill in these missing details (see "To the Student"). The writing has more of an engineering and science style to it than a formal mathematical format. Consequently, the arguments given are derivations in lieu of carefully constructed proofs. The exercises encourage students to think about the concepts and derivations rather than just grind out lots of routine solutions. The student who reads this book carefully and who solves most of the exercises will have a sound knowledge base to continue with a second-year partial differential equations course where careful proofs are constructed or with upper-division courses in science and engineering where detailed applications of partial differential equations are introduced. Both the exposition and exercises will build analytical skills that some students did not develop in the reform calculus courses.

The main changes in this second edition include a new section in Chapter 1 on advection and diffusion processes in the biological sciences, and a new chapter (Chapter 5) on models in the life sciences; the latter includes sections on age-structure, traveling epidemic waves, and pattern formation through chemical instabilities. Many new exercises have been added throughout the text.

In Chapter 1 we introduce some basic partial differential equations of applied mathematics. Many of the basic equations come from a conservation law, or balance law, and describe physical processes like advection (convection), diffusion, and reaction. The variety of applications show the central role that partial differential equations play in all areas of science and engineering. The goal is to give students a sense of the origins of partial differential equations and how their solutions differ. At the same time the exercises force students to revisit the chain rule, the divergence theorem, and other concepts from multivariable calculus.

Chapter 2 examines equations on unbounded domains, both infinite and semi-infinite. The author's view is that these problems are simpler than their counterparts on bounded domains with boundaries present. Most students have studied Laplace transforms in an elementary ordinary differential equations course, so it is a natural transition to study transform methods for partial differential equations.

A fundamental idea in applied mathematics is orthogonality. In Chapter 3, rather than adopt a strict focus on Fourier series, a general strategy is taken. Calculus courses have always included Taylor series, and many calculus courses, especially reform courses, now include some material on Fourier series. Therefore, students are ready to be introduced to general expansions of functions in series, especially orthogonal series. These expansions are motivated by the separation of variables method, and classical Fourier series arc studied as a special case.

Chapter 4 contains traditional material on the separation of variables method for solving partial differential equations on bounded domains. We solve equations with various boundary conditions in rectangular, cylindrical, and spherical geometry. Students are urged to use software packages to perform some of the calculations. There is a section on inverse problems and a section on the finite difference method.

I would like to greatly thank the many users of the first edition for helping make it a success. Its brevity certainly struck a positive chord. Some of their comments, corrections, and suggestions have become a part of this new edition. I would also like to thank my friend and bioscience colleague, Professor Tony Joern, for introducing me to a world of interesting problems in ecology. Some of these problems stimulated the new coverage of applications of PDEs to biology in this second edition.

Suggestions for Using the Text: The author has taught this material on numerous occasions and uses approximately the following schedule: Chapter 1 (10 classes), Chapter 2 (9 classes), Chapter 3 (7 classes), Chapter 4 (12 classes), Chapter 5 (6 classes). Under this schedule, the sections marked with an asterisk (*) in the Table of Contents are often not covered in lectures, but rather assigned as extra reading material to graduate students taking the course. Outlines of solutions to many of the exercises can be found at my web site (www.math.unl.edu/~dlogan) or accessed at the Springer-Verlag site (www.springer-ny.com).

Lincoln, Nebraska J. David Logan

To the Student

Partial differential equations (PDEs) is a subject about differential equations for unknown functions of several variables; the derivatives involved are partial derivatives. As such, it is a subject that is intimately connected with multivariable or third-semester calculus. To be successful you should have, first, a good command of the concepts in the calculus of several variables. So keep a calculus text nearby and review concepts when they are needed. The same comments apply to elementary ordinary differential equations (ODEs). There is an appendix at the end of the book that reviews some of the basic solution techniques for ODEs.

Second, a mathematics book should be read with a pencil and paper at hand. Elementary books fill in most of the steps in the exposition, but advanced books leave many details to the reader. This book has enough detail so that you can follow the discussion, but pencil and paper work is required in some portions. Verifying all the statements in a text is a worthwhile endeavor and will help you learn the material. Many students find that studying PDEs provides an opportunity to reinforce many calculus concepts and calculations.

Finally, the exercises are the most important part of this text, and you should try to solve most or all of them. Some will require routine analytical or computer calculations, but others will require careful thought. We learn mathematics by doing mathematics, even when we are stymied by a problem. The effort put into a failed attempt will help you sort out the concepts and reinforce the learning process. View the exercises as a challenge and resist the temptation to give up.

Lincoln, Nebraska

J. David Logan

Contents

CHAPTER

The Physical Origins of Partial Differential Equations

1.1 Mathematical Models

Many important ideas of mathematics were developed within the framework of physical science. For example, calculus has its origins in efforts to accurately describe the motion of bodies. Mathematical equations have always provided a language in which to formulate concepts in physics—Maxwell's equations describe electrodynamical phenomena, Newton's equations describe mechanical systems, Schrödinger's equation describes aspects of quantum mechanics, and so on. Over the years, however, mathematicians and scientists have extended these types of connections to include nearly all areas of science and technology, and a field has emerged called mathematical modeling. A *mathematical model* is an equation, or set of equations, whose solution describes the physical behavior of a related physical system. In this context, we say, for example, that Maxwell's equations are a model for electrodynamical phenomena. Like most mathematical models, Maxwell's equations are based on experiment and physical observations. In general, a mathematical model is a simplified description, or caricature, of physical reality expressed in mathematical terms. Mathematical modeling involves physical observation, selection of the relevant physical variables, formulation of the equations, analysis of the equations and simulation, and, finally, validation of the model. In this last step information from the simulations and solutions is fed back into the model to ascertain whether indeed the model does describe the phenomenon; at this step, modifications can occur and refinements can be made.

In this book we study partial differential equation (PDE) models. That is, we examine physical phenomena that can be described by equations called partial differential equations (PDEs). The focus is on the origin of such models and the tools used for their analysis.

The reader should be familiar with physical systems governed by ordinary differential equations (ODEs). For example, a typical ODE model in population ecology is the Malthus model (Thomas Malthus was an 18th century essayist who wrote on world population growth and resource supply)

$$\frac{du}{dt} = ru, \quad t > 0,$$

which is a simple model of population growth where the time rate of change of population $u = u(t)$ is proportional to the population. Here t is time, and $u = u(t)$ is the population of a given system of individuals. We refer to u as the state variable and say that the evolution of the state variable is governed by the model equation. The real number r is a given physical parameter that represents the relative growth rate; presumably, r could be measured for the given population under investigation. The solution to the model equation is easily found to be

$$u(t) = u_0 e^{rt}, \quad t > 0,$$

where u_0 represents the initial population. Thus, the Malthus model predicts exponential growth, which accurately describes some populations during beginning stages of growth.

The Malthus model is typical of ODE models. The state variable is a function of a single independent variable (time t), and the model contains a parameter, in this case r, that characterizes the physical system that is being modeled. In general, an ODE model has the form

$$\frac{du}{dt} = F(t, u; r), \quad t > 0,$$

where F is a given functional relation between t, u, and r. Often the model includes an initial condition of the form $u(0) = u_0$, where u_0 is a given state value at $t = 0$. More generally, an ODE model may consist of a system of n ODEs for n state variables $u_1(t), \ldots, u_n(t)$, and there can be m parameters $r_1, \ldots, r_m$.

A PDE model differs from an ODE model in that the state variable depends on more than one independent variable, and the resulting model equation is a PDE. Whereas an ODE models the evolution of a system in time, and observations are made in time, a PDE models the evolution of a system in both time and space; the system can be observed both in a time interval and in a spatial region (which may be one-, two-, or three-dimensional). PDE models may also be independent of time, but depend on several spatial variables. To fix the notion, let us consider the problem of determining the temperature in a laterally insulated metal bar of length l and unit cross-sectional area, whose two ends are maintained at a constant zero degrees and whose temperature initially (at time zero) varies along the bar and is given by a fixed function $\phi(x)$. See Figure 1.1.

How does the bar cool down? In this case, the state variable u is the temperature, and it depends upon both when the measurement is taken and where in the bar it is taken. Thus, $u = u(x, t)$, where t is time and $0 < x < l$. The model equation that governs the evolution of the temperature u is called the *heat equation* (we derive it in Section 1.3), and it has the form

$$u_t = ku_{xx}, \tag{1.1}$$

which is a PDE (we almost always use the subscript notation to indicate partial differentiation; and we rarely write the independent variables, preferring u to $u(x, t)$). In other words, the partial derivative of the temperature with respect to t must equal the second partial derivative of the temperature with respect to x, multiplied by a constant k. The constant k, called the diffusivity, is a known parameter and a property of the bar; it can be determined in terms of the density, specific heat, and thermal conductivity of the metal. So, the heat equation is a PDE model. Later we shall observe that it arises from a basic physical law (energy conservation) and an empirical observation (Fourier's heat law). The condition that the ends of the bar are maintained at zero degrees can be expressed by the equations

$$u(0, t) = 0, \quad u(l, t) = 0, \quad t > 0, \tag{1.2}$$

which are called *boundary conditions* because they impose conditions on the state variable on the boundary of the spatial domain. The condition that the bar initially has temperature $\phi(x)$ degrees is expressed

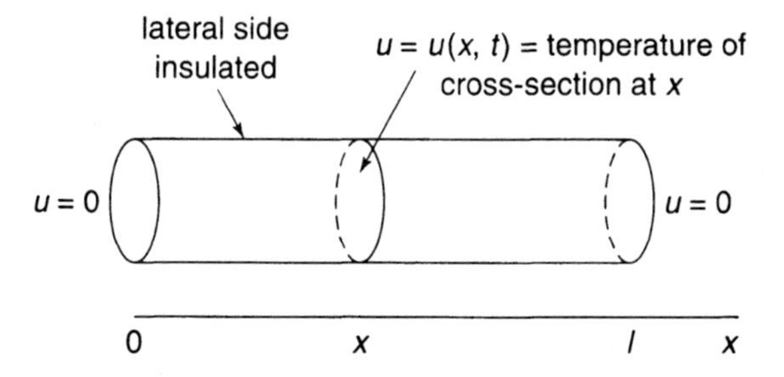

Figure 1.1. Metal bar laterally insulated with zero temperature at the ends. Heat flows in the x-direction, and $u(x, t)$ is the temperature of the cross-section at x at time t.

mathematically by

$$u(x, 0) = \phi(x), \quad 0 < x < l. \tag{1.3}$$

This condition is called an *initial condition* because it specifies the state variable at time $t = 0$. The entire set of equations (1.1)-(1.3)—the PDE and the auxiliary conditions—form the mathematical model for heat flow in the bar. Such a model in the subject of PDEs is called an *initial boundary value problem*. The invention and analysis of such models are the subjects of this book.

In the previous heat flow example, the state variable u depended upon two independent variables: a time variable and a spatial variable. Such a model is an evolution model. Some physical systems do not depend upon time, but rather only upon spatial variables. Such models are called *static*, *steady state*, or *equilibrium* models. For example, if Ω represents a bounded, three-dimensional spatial region in which no charges are present, and on the boundary $\partial\Omega$ of the region there is a given, time-independent electric potential (recall, in electrostatics, the gradient of the potential is the electric field vector), then it is known that the electric potential $u = u(x, y, z)$ inside Ω satisfies the Laplace equation, a partial differential equation having the form

$$u_{xx} + u_{yy} + u_{zz} = 0, \quad (x, y, z) \in \Omega. \tag{1.4}$$

If we denote the given boundary potential by $f(x, y, z)$, then (1.4) along with the *boundary condition*

$$u(x, y, z) = f(x, y, z), \quad (x, y, z) \in \partial\Omega, \tag{1.5}$$

is an equilibrium model for electrostatics. In PDEs such static models are called *boundary value problems*. Solving Laplace's equation (1.4) in a region Ω subject to a given condition (1.5) on the boundary is called the *Dirichlet problem*.

In general, a PDE in one spatial variable and time is an equation of the form

$$G(x, t, u, u_x, u_t, u_{xx}, u_{tt}, u_{xt}, \ldots) = 0, \ x \in \Omega, \quad t \in I, \tag{1.6}$$

where I is a given interval of time and Ω is an interval in one dimension. Often I is positive time $t \geq 0$, and Ω may be a bounded or unbounded interval. So, a PDE is an equation involving an unknown function $u = u(x, t)$, the state variable, and some of its partial derivatives. The *order* of equation (1.6) is the order of the highest derivative that occurs. A PDE model is a PDE supplemented with initial and/or boundary conditions that specify the state initially and on the boundary. One or more parameters, which are not explicitly shown, may also occur in (1.6).

PDEs are classified according to their order and to other properties. For example, as in ODEs, they are classified as linear or nonlinear. Equation (1.6) is *linear* if G is a linear function in u and in all of its derivatives; this means that the unknown u and the derivatives that are present appear alone and to the thc first power in the equation; otherwise, it is *nonlinear*. A linear equation is *homogeneous* if every term contains u or some derivative of u.

EXAMPLE
Both second-order equations $u_t + uu_{xx} = 0$ and $u_{tt} - u_x + \sin u = 0$ are nonlinear, the first because of the product uu_{xx} and the second because the unknown u is tied up in the nonlinear sine function. The second-order equation $u_t - \sin(x^2t)u_{xx} = 0$ is linear and homogeneous, and the first-order equation $u_t = -3xu_x + tx^2$ is linear and nonhomogeneous. □

One cannot overstate the significance of the partition of PDEs into the categories of linear and nonlinear. Linear equations have an algebraic structure to their solution sets; for example, the sum of two solutions to a homogeneous linear equation is again a solution, as are constant multiples of solutions. Another way of saying this is that solutions superimpose. Nonlinear equations do not share this property. Nonlinear equations are harder to solve, and their solution sets are more difficult to analyze. Even when nature presents us with a nonlinear model, we are likely to approximate it with a linear one. In this text we mostly study linear equations.

Equally important in classification schemes for a PDE is the specific nature of the physical phenomenon that it describes; for example, a PDEs can be classified as wave-like, diffusion-like, or static, depending upon whether it models wave propagation, a diffusion process, or an equilibrium state, respectively. For example, Laplace's equation (1.4) is a second-order, linear equilibrium equation; the heat equation (1.1) is a second-order, linear diffusion equation because heat flow is a diffusion process. In Section 1.10 we give a more precise characterization.

By a *solution* to the PDE (1.6) we mean a function $u = u(x, t)$ defined on the space–time domain $t \in I,\ x \in \Omega$ that satisfies, upon substitution, the equation (1.6) identically on that domain. Implicit in this definition is the stipulation that u possess as many continuous partial derivatives as required by the PDE. For example, a solution to a second-order equation should have two continuous partial derivatives so that it makes sense to calculate the derivatives and substitute them into the equation. Whereas the general solution to an ODE involves arbitrary constants, the general solution to a PDE involves arbitrary functions. Sometimes the general solution to a PDE can be found, but it is usually not necessary to have it to solve most problems of interest.

EXAMPLE

One should check, by direct substitution, that both functions

$$u_1(x, t) = x^2 + 2t \quad \text{and} \quad u_2(x, t) = e^{-t} \sin x$$

solve the heat equation

$$u_t - u_{xx} = 0.$$

There are many other solutions to this equation. Auxiliary conditions, like initial and boundary conditions, usually single out the appropriate solution to a problem. □

EXAMPLE

Consider the first-order, linear, nonhomogeneous PDE

$$u_x = t \sin x.$$

This equation can be solved by direct integration. We integrate with respect to x, holding t fixed, to get

$$u(x, t) = -t \cos x + \phi(t),$$

where ϕ is an arbitrary function. Notice that in PDEs, integration with respect to one variable produces an arbitrary function of the other variable, not an arbitrary constant as in one-dimensional calculus. This last equation defines the general solution. One can check that it is a solution for any differentiable function $\phi(t)$. PDEs have arbitrary functions in the expression for their general solutions; the number of such functions usually agrees with the order of the equation. □

At this point we mention that some computer algebra systems have commands that return the general solution to a PDE in terms of arbitrary functions. The reader who wishes to explore this feature in the Maple software package may at this time go directly to Section 2.8 for examples.

Geometrically, a solution $u = u(x, t)$ can be thought of as a surface in xtu-space. Refer to Figure 1.2. The surface lies over the space–time domain $\Omega \times I$ (this is the set (x, t) such that $x \in \Omega$ and $t \in I$). Alternatively, one can regard the solution as a continuous sequence of time snapshots. That is, for each fixed time t_0, $u(x, t_0)$ is a function of x and thus represents a time snapshot of the solution. In different words, $u(x, t_0)$ is the trace of the solution surface $u = u(x, t)$ taken in the $t = t_0$-plane. In some contexts, $u(x, t_0)$ is interpreted as a wave profile, or signal, at time t_0. In this way a solution $u(x, t)$ of (1.6) can be regarded a continuous sequence, in time, of evolving wave forms.

Bibliographic Notes

There are many excellent elementary PDE books written at about the same level as this one. We mention Farlow (1993), Guenther & Lee (1992),

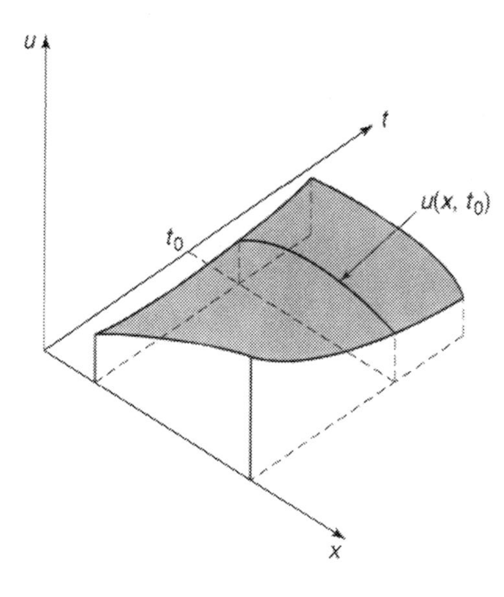

Figure 1.2. A solution surface $u = u(x, t)$ whose cross-section is a wave profile.

and Strauss (1992). At an advanced level we suggest John (1982) or Renardy & Rogers (1993). The classic text by Tychonov & Samarskii (1990) has a good balance of applications and theory. Nonlinear PDEs are treated in detail in Logan (1994), Smoller (1995), and Whitham (1974). PDE models occur in every area of the pure and applied sciences. General texts involving modeling are Lin & Segel (1989) and Logan (1997). Specific areas are covered in Bird, Stewart, & Lightfoot (1960) (chemical engineering), Carslaw & Jaeger (1959) (heat transfer), Chorin & Marsden (1993) (fluid dynamics), Edelstein-Keshet (1988), Grindrod (1997), Kot (2001), Britton (2003), and Murray (2003) (biology), de Marsily (1987) and Logan (2001) (hydrogeology), and Segel (1987) (continuum mechanics).

Exercises

1. Verify that a solution to the heat equation (1.1) on the domain $-\infty < x < \infty,\ \ t > 0$ is given by

$$u(x, t) = \frac{1}{\sqrt{4\pi kt}} e^{-x^2/4kt}.$$

Use a computer algebra package to sketch several time snapshots on the same set of coordinate axes to show how the temperature profile evolves in time (take $k = 1$). What do the temperature profiles look like as $t \to 0$? Sketch the solution surface. How does the parameter k affect the solution?

2. Verify that $u(x, y) = \ln \sqrt{x^2 + y^2}$ satisfies the Laplace equation

$$u_{xx} + u_{yy} = 0$$

for all $(x, y) \neq (0, 0)$.

3. Find a function $u = u(x, t)$ that satisfies the PDE

$$u_{xx} = 0, \quad 0 < x < 1, \ t > 0,$$

subject to the boundary conditions

$$u(0, t) = t^2, \quad u(1, t) = 1, \quad t > 0.$$

4. Verify that

$$u(x, t) = \frac{1}{2c} \int_{x-ct}^{x+ct} g(\xi) d\xi$$

is a solution to the PDE $u_{tt} = c^2 u_{xx}$, where c is a constant and g is a continuously differentiable function. Here you will need to use Leibniz's rule for differentiating an integral with respect to a parameter that occurs in the limits of integration and in the integrand; Leibniz's rule is

$$\frac{d}{dt} \int_{a(t)}^{b(t)} F(y, t) dy = \int_{a(t)}^{b(t)} F_t(y, t) dy + F(b(t), t) b'(t) - F(a(t), t) a'(t).$$

Here, assume that a, b, and F are continuously differentiable.

5. For what values of a and b will $u(x, t) = e^{at} \sin bx$ solve the heat equation

$$u_t - k u_{xx} = 0?$$

6. Find the general solution to the equation $u_{xt} + 3u_x = 1$. Hint: Let $v = u_x$ and solve the resulting equation for v; then find u.

7. Show that the nonlinear equation $u_t = u_x^2 + u_{xx}$ can be reduced to the heat equation (1.1) by changing the dependent variable via $w = e^u$.

8. Show that the function $u(x, y) = \arctan(y/x)$ satisfies the two-dimensional Laplace's equation $u_{xx} + u_{yy} = 0$ for $y > 0$. Using this fact, construct a solution to the two-dimensional Laplace's equation in $y > 0$ that also satisfies the boundary conditions $u(x, 0) = 1$ for $x > 0$ and $u(x, 0) = -1$ for $x < 0$. Be sure to explain which branch of the arctan function you are using.

9. Show that $e^{-\xi y} \sin(\xi x)$, $x \in \mathbf{R}$, $y > 0$, is a solution to $u_{xx} + u_{yy} = 0$ for any value of the parameter ξ. Deduce that

$$u(x, y) = \int_0^\infty c(\xi) e^{-\xi y} \sin(\xi x) d\xi$$

is a solution to the same equation for any function $c(\xi)$ that is bounded and continuous on $[0, \infty)$. (The hypotheses on c allow you to differentiate under the integral sign.) This exercise shows that taking integrals of solutions sometimes gives another solution; integration is a way of superimposing, or adding, a continuum of solutions.

10. Linear, homogeneous PDEs with constant coefficients admit complex solutions of the form $u(x, t) = Ae^{i(kx - \omega t)}$, which are called *plane waves*. The real and imaginary parts of this complex function give real solutions. Here, A is the amplitude, k is the wave number, and ω is the temporal frequency.

When the plane wave form is substituted into a PDE there results a *dispersion relation* of the form $\omega = \omega(k)$, which states how the frequency depends upon the wave number. For the following PDEs, find the dispersion relation and describe the resulting plane wave by sketching wave profiles at different times.

(a) $u_t = Du_{xx}$.
(b) $u_{tt} - c^2 u_{xx} = 0$.
(c) $u_t + u_{xxx} = 0$.
(d) $u_t = iu_{xx}$.

1.2 Conservation Laws

Many PDE models come from a basic balance law, or conservation law. A conservation law is just a mathematical formulation of the basic fact that the rate at which a quantity changes in a given domain must equal the rate at which the quantity flows across the boundary plus the rate at which the quantity is created, or destroyed, within the domain. For example, consider a population of a certain animal species in a fixed geographical region. The rate of change of the animal population must equal the rate at which animals migrate into the region, minus the rate at which they migrate out, plus the birth rate, minus the death rate. Such a statement is a verbal expression of a balance, or conservation, law. One can make similar kinds of statements for many quantities—heat energy, the mass of a chemical, the number of automobiles on a freeway, and so on.

To quantify such statements we require some notation. Let the state variable $u = u(x, t)$ denote the density of a given quantity (mass, energy, animals, automobiles, etc.); density is usually measured in amount per unit volume, or sometimes amount per unit length. For example, energy density is measured in energy units per volume. We assume that any variation in the quantity be restricted to one spatial dimension, that is, we assume a one-dimensional domain (say, a tube, as in Figure 1.3) where each cross-section is labeled by the spatial variable x; we require that

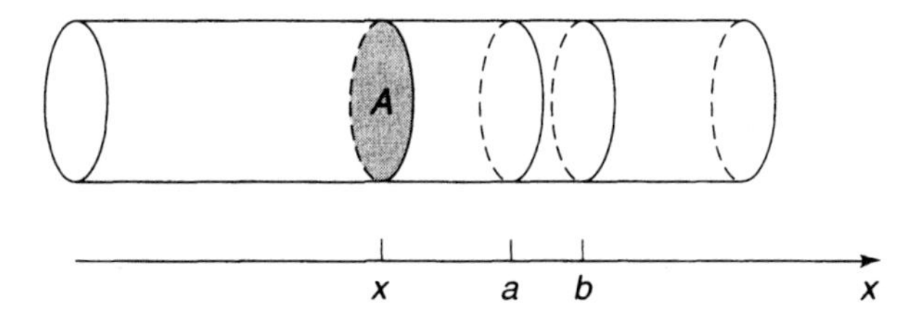

Figure 1.3. Tube with cross-sectional area A shown with arbitrary section. The lateral sides are insulated, and quantities vary only in the x-direction and in time.

there be no variation of $u(x, t)$ within the cross-section at x. Implicit is the assumption that the quantity in the tube is abundant and continuous enough in x so that it makes sense to define a *density* at each section of the tube. The amount of the quantity in a small section of width dx is $u(x, t)Adx$, where A is the cross-sectional area of the tube. Further, we let $\phi = \phi(x, t)$ denote the *flux* of the quantity at x, at time t. The flux measures the amount of the quantity crossing the section at x at time t, and its units are given in amount per unit area, per unit time. Thus, $A\phi(x, t)$ is the actual amount of the quantity that is crossing the section at x at time t. By convention, flux is positive if the flow is to the right, and negative if the flow is to the left. Finally, let $f = f(x, t)$ denote the given rate at which the quantity is created, or destroyed, within the section at x at time t. The function f is called a source term if it is positive, and a sink if it is negative; it is measured in amount per unit volume per unit time. Thus, $f(x, t)Adx$ represents the amount of the quantity that is created in a small width dx per unit time.

A conservation law is a quantitative relation between u, ϕ, and f. We can formulate the law by considering a fixed, but arbitrary, section $a \le x \le b$ of the tube (Figure 1.3) and requiring that the rate of change of the total amount of the quantity in the section must equal the rate at which it flows in at $x = a$, minus the rate at which it flows out at $x = b$, plus the rate at which it is created within $a \le x \le b$. In mathematical symbols,

$$\frac{d}{dt}\int_a^b u(x, t)Adx = A\phi(a, t) - A\phi(b, t) + \int_a^b f(x, t)Adx. \tag{1.7}$$

This equation is the fundamental conservation law; it is an integral expression of the basic fact that there must be a balance between how much goes in, how much goes out, and how much is changed. Because A is constant, it may be canceled from the formula.

Equation (1.7) is an integral model. However, if the functions u and ϕ are sufficiently smooth, then it may be reformulated as a PDE model. For example, if u has continuous first partial derivatives, then the time derivative on the left side of (1.7) may be brought under the integral sign to obtain

$$\frac{d}{dt}\int_a^b u(x, t)dx = \int_a^b u_t(x, t)dx$$

(note that this is a special case of Leibniz's rule—see Section 1.1, Exercise 4). If ϕ has continuous first partials, then the fundamental theorem of calculus can be applied to write the change in flux as the integral of a derivative, or

$$\phi(a, t) - \phi(b, t) = -\int_a^b \phi_x(x, t)dx.$$

Therefore, (1.7) may be written

$$\int_a^b (u_t(x,t) + \phi_x(x,t) - f(x,t))dx = 0.$$

Because $a \leq x \leq b$ can be any interval whatsoever, and because the integrand is continuous, it follows that the integrand must vanish identically, or

$$u_t(x,t) + \phi_x(x,t) = f(x,t). \tag{1.8}$$

Equation (1.8) is a local version of (1.7), obtained under the assumption that u and ϕ are continuously differentiable; it is a PDE model describing the relation between the density, its flux, and the rate at which it is created. We shall call the PDE (1.8) the *fundamental conservation law*. The f-term is called the source term, and the ϕ-term is called the flux term. In (1.7) we usually drop the understood notational dependence on x and t and just write $u_t + \phi_x = f$ for simplicity.

Before studying some examples, we make some general comments. The flux ϕ and source f are functions of x and t, but their dependence on x and t may be through dependence upon the density u itself. For example, the source term f may be given as a function of density via $f = f(u)$, where, of course, $u = u(x,t)$. Similarly, ϕ may depend on u. These dependencies may lead to a nonlinear model. Next, we observe that (1.8) is a single equation, yet there are two unknowns, u and ϕ (usually the source f is assumed to be given). This means that another equation is required, which relates u and ϕ. Such equations are called *constitutive relations* (or equations of state), and they arise from physical assumptions about the medium.

Example

(*Advection*) A model where the flux is proportional to the density itself, that is,

$$\phi = cu,$$

where c is a constant, is called a *advection* model. Notice that c must have velocity units (length per time). In this case the conservation law (1.8) becomes, in the absence of sources ($f = 0$),

$$u_t + cu_x = 0. \tag{1.9}$$

Equation (1.9) is called the *advection equation*. The reader should verify that the function

$$u(x,t) = F(x - ct) \tag{1.10}$$

is a solution to (1.9) for any differentiable function F. Such solutions (1.10) are called *right-traveling waves*, since the graph of $F(x - ct)$ is just the graph of $F(x)$ shifted to the right ct spatial units. So, as time t increases, the wave profile $F(x)$ moves to the right, undistorted, with its shape unchanged, at speed c. Figure 1.4 shows two ways of viewing a traveling wave. Intuitively, (1.9) describes what we usually call advection. For example, a swarm of insects moving as an undistorted density wave would represent advection. Other common descriptive terms for this kind of movement are *transport* and *convection*. □

If the flux is a nonlinear function of the density, that is, $\phi = \phi(u)$, then the conservation law (1.8) (again take $f = 0$) takes the form

$$u_t + \phi(u)_x = u_t + \phi'(u)u_x = 0. \tag{1.11}$$

If $\phi(u)$ is not linear in u, then (1.11) is a model of *nonlinear* advection, and such models are more difficult to analyze. For the most part, we do not study such models in this text. The reader can consult Logan (1994) for an elementary treatment of nonlinear equations.

Example

(*Advection and decay*) Recall from elementary differential equations that decay (e.g., radioactive decay) is modeled by the law $du/dt = -\lambda u$, where λ is the decay rate. Thus, a substance advecting through a tube at velocity c (for example, a radioactive chemical dissolved in water flowing at speed c) would be modeled by the advection–decay equation

$$u_t + cu_x = -\lambda u. \tag{1.12}$$

Here, $f = -\lambda u$ is the source term (specifically, the decay term) and $\phi = cu$ is the flux term in the conservation law (1.8). □

The pure initial value problem for the advection equation is

$$u_t + cu_x = 0, \quad x \in \mathbf{R}, t > 0, \tag{1.13}$$

$$u(x, 0) = u_0(x), \quad x \in \mathbf{R}, \tag{1.14}$$

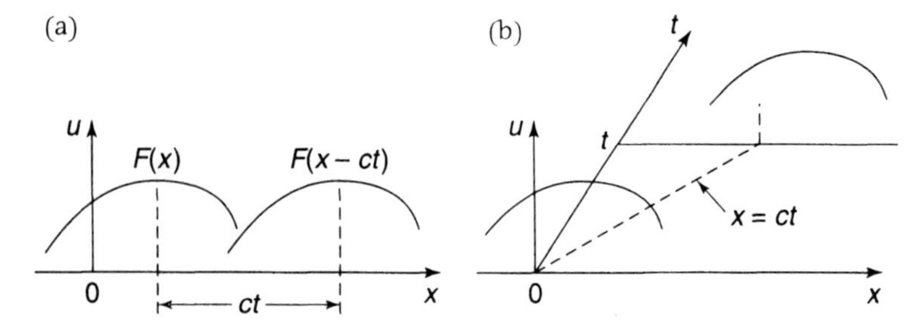

Figure 1.4. Two views of a traveling wave: (a) time snapshots, and (b) in space–time as a ridge.

where $u_0(x)$ is a given initial density, or signal. From (1.10) it follows that the solution to this problem (1.13)–(1.14) is

$$u(x,t) = u_0(x - ct).$$

So, physically, the initial density signal moves to the right at speed c. Alternatively, we can think of the density signal moving along the family of straight lines $\xi = x - ct =$ constant in space–time. These lines, called *characteristics*, are the curves that carry the signal.

Now let us indicate how to solve a general advection equation of the form

$$u_t + cu_x + au = f(x,t), \tag{1.15}$$

where a and c are constants and f is a given function. Because the advection equation propagates signals at speed c, it is reasonable to attempt to solve this equation by transforming to a new, moving, coordinate system. Thus, let ξ and τ be new independent variables, called *characteristic coordinates*, defined by

$$\xi = x - ct, \quad \tau = t.$$

We think of ξ as a moving coordinate that travels (advects) with the signal. Then, if we denote $u(x,t)$ in the new variables by $U(\xi,\tau)$ (that is, $U(\xi,\tau) = u(\xi + c\tau, \tau)$, or $u(x,t) = U(x - ct, t)$), then by the chain rule,

$$u_t = U_\xi \xi_t + U_\tau \tau_t = -cU_\xi + U_\tau$$

and

$$u_x = U_\xi \xi_x + U_\tau \tau_x = U_\xi.$$

So equation (1.15) becomes simply

$$U_\tau + aU = F(\xi,\tau),$$

where $F(\xi,\tau) = f(\xi + c\tau, \tau)$. Observe that a PDE like this one that contains derivatives with respect to *only one* of its independent variables can be regarded as an ODE with the other independent variable as a parameter. Therefore it can be solved by ODE methods, which are reviewed in the Appendix. It has the form of a linear equation, and so it can be solved by multiplying by the integrating factor $e^{a\tau}$ and integrating with respect to τ. An example illustrates this procedure.

Example

Find the general solution of

$$u_t + 2u_x - u = t.$$

Let $\xi = x - 2t$, $\tau = t$. In these characteristic coordinates the equation becomes

$$U_\tau - U = \tau.$$

Multiplying by $e^{-\tau}$ gives

$$\frac{\partial}{\partial \tau}(Ue^{-\tau}) = \tau e^{-\tau}.$$

Integrating,

$$Ue^{-\tau} = \int \tau e^{-\tau} d\tau = -(1+\tau)e^{-\tau} + g(\xi),$$

where g is an arbitrary function. Transforming back to xt variables then gives the general solution

$$u(x,t) = -(1+t) + g(x-2t)e^{t}.$$

□

The more general reaction–advection PDE

$$u_t + cu_x = f(x,t,u)$$

can theoretically be solved by making the same transformation $\xi = x - ct, \quad \tau = t$ to turn it into a simpler equation of the form

$$U_\tau = F(\xi, \tau, U).$$

In these characteristic coordinates the PDE simplifies to the form of an ODE with only one derivative. The important point is that the advection operator $\frac{\partial}{\partial t} + c\frac{\partial}{\partial x}$ simplifies to $\frac{\partial}{\partial \tau}$ in characteristic coordinates; thus, changing independent variables this way is a good strategy for handling equations having advection operators. This solution technique is called the *method of characteristics*.

A similar characteristic method can be applied to solve the equation

$$u_t + c(x,t)u_x = f(x,t,u).$$

In this case, we think of $c(x,t)$ as the advection speed in a heterogeneous medium; it replaces the constant c in the previous problem and now depends on the location in the medium and on time. The characteristic coordinates are given by $\xi = \phi(x,t), \quad \tau = t$, where $\phi(x,t) = C$ is the general solution of the ODE

$$\frac{dx}{dt} = c(x,t).$$

In these new coordinates it is easy to see that the original PDE transforms into an equation of the form

$$U_\tau = F(\xi, \tau, U),$$

where $U = U(\xi, \tau)$.

Example

Consider the PDE

$$u_t + 2tu_x = 0.$$

Here, $c(x, t) = 2t$. Setting $\frac{dx}{dt} = 2t$ and solving gives $x - t^2 = C$. Thus, $\xi = \phi(x, t) = x - t^2$. The characteristic coordinates are

$$\xi = x - t^2, \quad \tau = t,$$

and we find by the chain rule that

$$u_t = U_\xi(-2t) + U_\tau, \quad u_x = U_\xi.$$

Therefore $u_t + 2tu_x = U_\tau$ and the original PDE transforms into $U_\tau = 0$. Hence $U = g(\xi)$, where g is an arbitrary function. The general solution to the given PDE is thus $u(x, t) = g(x - t^2)$. Observe that the solution is constant along the set of characteristic curves (parabolas in space–time) $x - t^2 = C$. □

In Section 1.4 there is an expanded treatment of advection in a biological context.

Exercises

1. How does the basic conservation law (1.8) change if the tube has variable cross-sectional area $A = A(x)$ rather than a constant cross-sectional area?
2. Solve the initial value problem

$$u_t + cu_x = 0, \quad x \in \mathbf{R},\ t > 0; \quad u(x, 0) = e^{-x^2}, \ x \in \mathbf{R}.$$

   Pick $c = 2$ and sketch the solution surface and several time snapshots. Do you see a traveling wave? Sketch the characteristic curves in the xt-plane.
3. Find the general solution of the advection–decay equation (1.12) by transforming to characteristic coordinates $\xi = x - ct, \quad \tau = t$.
4. Show that the decay term in the advection–decay equation (1.12) can be removed by making a change of the dependent variable to $w = ue^{\lambda t}$.
5. Solve the pure initial value problems

$$u_t + xtu_x = 0, \quad u(x, 0) = f(x)$$

   and

$$u_t + xu_x = e^t, \quad u(x, 0) = f(x).$$

6. Solve the initial boundary value problem

$$u_t + cu_x = -\lambda u, \quad x, t > 0,$$

$$u(x, 0) = 0, \quad x > 0; \qquad u(0, t) = g(t), \quad t > 0.$$

   In this problem you will have to treat the domains $x > ct$ and $x < ct$ differently; the boundary condition affects the solution region $x < ct$, and the initial condition affects it in the region $x > ct$.

7. Solve the pure initial value problem

$$u_t + u_x - 3u = t, \quad x \in \mathbf{R}, \ t > 0,$$

$$u(x, 0) = x^2, \quad x \in \mathbf{R}.$$

8. To study the absorption of nutrients in an insect gut we model its digestive tract by a tube of length l and cross-sectional area A. Nutrients of concentration $n = n(x, t)$ flow through the tract at speed c, and they are adsorbed locally at a rate proportional to $\sqrt{n}$. What is the PDE model? If the tract is empty at $t = 0$ and then nutrients are introduced at the constant concentration n_0 at the mouth ($x = 0$) for $t > 0$, formulate an initial boundary value problem for $n = n(x, t)$. Solve this PDE model and sketch a graph of the nutrient concentration exiting the tract (at $x = l$) for $t > 0$. Physically, why is the solution $n(x, t) = 0$ for $x > ct$?

9. Explain why the function $u(x, t) = G(x + ct)$ is called a left-traveling wave. How would you attempt to solve the advection equation $u_t - cu_x = F(x, t, u)$?

10. The density of cars on a busy one-lane freeway with no exits and entrances is $u = u(x, t)$ cars per mile. If $\phi = \phi(x, t)$ is the flux of cars, measured in cars per hour, derive a conservation law relating the density and flux. Why would $\phi = \alpha u(\beta - u)$ $(\alpha, \beta > 0)$ be a reasonable assumption? Write down the resulting nonlinear PDE for u.

11. Find a formula that implicitly defines the solution $u = u(x, t)$ of the initial value problem for the reaction–advection equation

$$u_t + vu_x = -\frac{\alpha u}{\beta + u}, \quad x \in \mathbf{R}, \ t > 0,$$

$$u(x, 0) = f(x), \quad x \in \mathbf{R}.$$

Here, v, α, and β are positive constants. Show from the implicit formula that you can always solve for u in terms of x and t.

12. Write a formula for the general solution of the equation

$$u_t + cu_x = f(x)u$$

1.3 Diffusion

Let us again write down the basic conservation law (1.8) with no sources:

$$u_t + \phi_x = 0. \tag{1.16}$$

To reiterate, $u = u(x, t)$ represents the density of a physical quantity, and $\phi = \phi(x, t)$ represents its flux. Equation (1.16) describes locally how changes in density are related to changes in flux. In the last section we modeled advection by assuming that the flux was proportional to the density (flux equals velocity times density). Now we want to model a simple diffusion process. To fix the notion let u denote the concentration of some chemical species, say a gas, in a tube. We expect that the random motion and collisions of the molecules will cause concentrations of the gas to spread out; the gas will move from higher concentrations to lower concentrations. The same could be said for insects in a tube, people congregated in a hallway, or heat energy in a metal bar.

To model this type of random motion we make two observations: (i) the movement is from higher concentrations to lower concentrations, and (ii) the steeper the concentration gradient, the greater the flux. Therefore, the flux should depend on the x-derivative of the density (which measures the steepness of the density curve). Assuming a simple linear relationship, we take

$$\phi = -Du_x, \tag{1.17}$$

where D is a constant of proportionality. The minus sign guarantees that if $u_x < 0$, then ϕ will be positive and the flow will be, by our convention, to the right; if $u_x > 0$, then ϕ will be negative and the flow will be to the left. We say that the flow is *down the gradient*. Equation (1.17) is called *Fick's law*, and the constant D is called the *diffusion constant*; D is measured in length-squared per unit time.

When the constitutive equation (1.17) is substituted into the conservation law (1.16), we obtain a simple model equation

$$u_t - Du_{xx} = 0, \tag{1.18}$$

which is called the *diffusion equation*. This PDE model is one of the fundamental equations in applied mathematics.

EXAMPLE

(*Heat Flow*) Let us consider heat flow in a one-dimensional bar having a constant density ρ and constant specific heat C. Both of these constants are physical parameters that are tabulated in engineering and physics handbooks. The specific heat is the amount of energy required to raise a unit mass of material one degree, and it is given in units of energy per mass per degree. We may apply the basic conservation law (1.16) to the bar, with u being the energy density given by $u(x,t) = \rho C\theta(x,t)$, where $\theta = \theta(x,t)$ is the temperature at (x,t) (the stipulation that energy is proportional to temperature is, in itself, an assumption about the medium). Therefore,

$$\rho C\theta_t + \phi_x = 0 \tag{1.19}$$

is an expression of energy balance in the bar when no sources are present. The energy flux ϕ is assumed to be given by a Fick's law-type expression

$$\phi = -K\theta_x, \tag{1.20}$$

where K is thermal conductivity, another physical constant. In the context of heat flow, (1.20) is called the *Fourier heat law*. This law, a constitutive relation based on empirical evidence, is a statement of the fact that heat flows from hotter regions to colder regions; stated differently, heat flows down the temperature gradient. Now, we may substitute (1.20) into (1.19) to obtain a single equation for the temperature $\theta(x, t)$, namely,

$$\theta_t - k\theta_{xx} = 0, \quad k \equiv \frac{K}{\rho C}. \tag{1.21}$$

Equation (1.21) is the *heat equation*; it is the diffusion equation in the context of heat flow. The constant k is called the diffusivity and is a property of the medium; values of k for different media (metals, plastics, etc.) can be found in physical handbooks. Note that k has the same dimensions (length-squared per time) as the diffusion constant. In the sequel we shall often use u in place of θ for the temperature function. □

In some cases the thermal conductivity K in (1.20) may not be constant, but rather may depend on x if the bar is nonhomogeneous; over large temperature ranges the conductivity could also depend on the temperature θ. If, for example, $K = K(\theta)$, then we obtain a nonlinear heat model

$$\rho C\theta_t - (K(\theta)\theta_x)_x = 0.$$

It is possible, of course, that the density and specific heat could depend on the location x or the temperature θ.

EXAMPLE

(*Advection-Diffusion*) If both diffusion and advection are present, then the flux is given by

$$\phi = cu - Du_x,$$

and the conservation law (1.16) becomes

$$u_t + cu_x - Du_{xx} = 0,$$

which is the *advection-diffusion equation*. This equation would govern the density of a chemical, say, that is being advected by the bulk motion of a fluid moving at velocity c in which it is dissolved, while at the same time it is diffusing according to Fick's law. If the chemical also decays at rate λ, then we include a source term, and the model is

$$u_t + cu_x - Du_{xx} = -\lambda u,$$

which is a advection-diffusion-decay equation. □

Diffusion-like equations are often accompanied by an initial condition that specifies the density at time $t = 0$ at all points of the spatial domain, as well as boundary conditions that specify conditions on the density at the boundaries of the domain for all time. To fix the idea, let us consider a tube of finite length $0 \le x \le l$. Then an initial condition has the form

$$u(x, 0) = u_0(x), \quad 0 \le x \le l,$$

where u_0 is the given initial density distribution. There are three types of boundary conditions that often occur in physical problems. If the density is specified at the boundary $x = 0$, then we have

$$u(0, t) = g(t), \quad t > 0,$$

where g is given; this condition is called a *Dirichlet condition*. We could also specify the flux at a boundary, i.e.,

$$-Du_x(0, t) = h(t), \quad t > 0,$$

which is called a *Neumann condition*. If $h(t) \equiv 0$, then we say that the boundary is *insulated*; if g or h is zero, we also say that the corresponding boundary condition is *homogeneous*.

A third type of boundary condition has the form (say, at $x = 0$)

$$-Du_x(0, t) = -\beta(u(0, t) - \psi(t)), \quad t > 0.$$

In heat flow in a bar, for example, this law expresses Newton's law of cooling, which states that the heat flux is proportional to the temperature difference between the end of the bar and the given temperature $\psi(t)$ of the environment; β is the constant of proportionality representing a heat-loss factor. These types of conditions are also called *radiation* conditions, or *Robin* conditions.

Part of the requirement of a well-posed PDE model is that the PDE be accompanied by appropriate initial and boundary conditions so that the resulting mathematical problem has a unique solution as well as a correct physical meaning.

Many PDE models have the property that after a long time the transients caused by initial conditions decay away and the solution approaches a *steady-state* $u = u(x)$, dependent only upon x and the boundary conditions. In this case the time-derivatives in the equation vanish and $u(x)$ satisfies an ODE with the appropriate boundary conditions. A simple example is given by the diffusion equation $u_t = Du_{xx}$. The steady-state solution $u = u(x)$ satisfies $Du''(x) = 0$, which gives a linear density profile $u(x) = ax + b$. The constants a and b can be determined by the given boundary conditions. The following example is more involved.

EXAMPLE
Consider an initial boundary value problem for the diffusion-decay equation:

$$u_t = Du_{xx} - ru, \quad 0 < x < L, \quad t > 0,$$
$$u(0,t) = 0, \quad -Du_x(L,t) = -1, \quad t > 0,$$
$$u(x,0) = g(x), \quad 0 < x < L.$$

This model could, for example, represent a diffusing fish population in a canal with *per capita* death rate r. At the left boundary the population density is maintained at zero, while at the right boundary fish are introduced into the canal at rate 1 per unit area per unit time. The steady-state model is the pure boundary value problem

$$Du'' - ru = 0, \quad 0 < x < L,$$
$$u(0) = 0, \quad -Du'(L) = -1$$

for $u = u(x)$. The initial condition is ignored since it no longer affects the solution. To solve the differential equation we apply elementary methods from Appendix A to obtain

$$u(x) = c_1 \sinh(\sqrt{r/D}x) + c_2 \cosh\sqrt{(r/Dx)},$$

where c_1 and c_2 are arbitrary constants. The boundary condition $u(0) = 0$ forces $c_2 = 0$, giving $u(x) = c_1 \sinh(\sqrt{r/D}x)$. The right boundary condition gives $-D\sqrt{r/D}c_1 \cosh(\sqrt{r/D}L) = -1$, or $c_1 = 1/\sqrt{rD}\cosh(\sqrt{r/D}L)$. Therefore the steady-state solution, representing the long time density of fish in the canal, is

$$u(x) = \frac{\sinh(\sqrt{r/D}x)}{\sqrt{rD}\cosh(\sqrt{r/D}L)}.$$

□

Pure boundary value problems for ODEs are unlike initial value problems in that they do not always have a solution, or they may have many solutions. Thus, there is no guarantee that the system has a steady-state solution. If there is a steady-state, it may be unstable and the system will not approach that state. These issues warrant further analysis, and they are discussed in Chapters 4 and 5.

Exercises

1. Heat flows longitudinally through a metal bar of length 10 centimeters, and the temperature $u = u(x,t)$ satisfies the diffusion equation $u_t = ku_{xx}$, where $k = 0.02$ square centimeters per second. Suppose the temperatures at some fixed time T at $x = 4, 6, 8$ cm are 58, 64, and 72 degrees, respectively. Estimate

$u_{xx}(6, T)$ using a difference approximation. Will the temperature at $x = 6$ increase or decrease in the next instant of time? Estimate the temperature at $x = 6$ at $T + 0.5$ seconds. (Recall from calculus that

$$f''(a) \approx \frac{f(a-h) - 2f(a) + f(a+h)}{h^2},$$

where h is a small increment.)

2. Let $u = u(x,t)$ satisfy the PDE model

$$u_t = ku_{xx}, \qquad 0 < x < l, \quad t > 0,$$
$$u(0,t) = u(l,t) = 0, \qquad t > 0,$$
$$u(x,0) = u_0(x), \qquad 0 \le x \le l.$$

Show that

$$\int_0^l u(x,t)^2 dx \le \int_0^l u_0(x)^2 dx, \quad t \ge 0.$$

Hint: Let $E(t) = \int_0^l u(x,t)^2 dx$ and show that $E'(t) \le 0$. What can be said about $u(x,t)$ if $u_0(x) = 0$?

3. Show that the problem

$$u_t = ku_{xx}, \qquad 0 < x < l, \quad t > 0,$$
$$u(0,t) = g(t), \qquad u(l,t) = h(t), \quad t > 0,$$
$$u(x,0) = u_0(x), \qquad 0 \le x \le l,$$

with nonhomogeneous boundary conditions can be transformed into a problem with homogeneous boundary conditions. Hint: Introduce a new dependent variable w by subtracting from u a linear function of x that satisfies the boundary conditions at any fixed t. In the transformed problem for w, notice that the PDE picks up a source term, so you are really trading boundary conditions for source terms.

4. Show that the advection–diffusion–decay equation

$$u_t = Du_{xx} - cu_x - \lambda u$$

can be transformed into the diffusion equation by a transformation of the form

$$u(x,t) = w(x,t)e^{\alpha x - \beta t}.$$

Solution: Take $\alpha = c/(2D)$, $\beta = \lambda + c^2/(4D)$.

5. Heat flow in a metal rod with an internal heat source is governed by the problem

$$u_t = ku_{xx} + 1, \quad 0 < x < 1, \quad t > 0,$$
$$u(0,t) = 0, \quad u(1,t) = 1, \quad t > 0.$$

What will be the steady-state temperature in the bar after a long time? Does it matter that no initial condition is given? Hint: In the steady state u depends only on x.

6. A bar loses heat across its lateral boundary at a rate proportional to the temperature u. The equation is

$$u_t = ku_{xx} - au, \quad 0 < x < 1, \; t > 0,$$

$$u(0,t) = 1, \quad u(1,t) = 1, \quad t > 0.$$

Graph the steady-state temperature distribution in the bar and analyze how heat is flowing in the bar and through its boundaries.

7. Bacteria in a one-dimensional medium (a tube of unit cross-sectional area, length l, and capped on both ends) have a growth rate given by the logistic law $ru(1 - u/K)$, where r is the growth constant, K is the carrying capacity, and $u = u(x,t)$ is the density of bacteria measured in bacteria per unit length. Initially, the density is given by $u = ax(l - x)$. For $t > 0$ the bacteria also diffuse, with diffusion constant D. Formulate an initial boundary value problem for the density. If we wait a long time, what will the density be? Use your intuition to sketch a sequence of density profiles that show how the density evolves. You may want to consider the cases $al^2 < 4K$ and $al^2 > 4K$ separately. Hint: Think about the problem if no diffusion were present.

8. Show that the equation $u_t = k(t)u_{xx}$ can be transformed into the diffusion equation by changing the independent variable time t to $\tau = \int_0^t k(\eta)d\eta$. Show that the equation $u_t = ku_{xx} - b(t)u_x$ can be transformed into the diffusion equation by changing the spatial variable to $\xi = x - \int_0^t b(\eta)d\eta$.

1.4 PDE Models in Biology

In many ways mathematical models in biology are quite different from those in the physical sciences and engineering. In the latter case models are usually based on scientific principles and laws that can be expressed precisely in terms of nearly exactly measurable quantities, and the goal is to obtain precise quantitative results. For example, there is little question that Newton's laws of motion or Maxwell's equations of electrodynamics might apply to some physical situation. In the life sciences, however, it is really impossible to model all of the intricacies of predator–prey interactions, tumor growth dynamics, or the spread of an infectious disease. Therefore, the models are more phenomenological in nature, predicting only qualitative features rather than leading to detailed quantitative results. Biological systems are highly complex, and often we attempt to describe only the gross features, ignoring much of the fine detail, the stochasticity, and the natural variability. In this section we introduce some standard PDE models in the biological sciences. The ideas will complement the discussion of advection and diffusion introduced in the previous two sections.

We begin with a simple thought experiment that illustrates how PDEs arise naturally in population dynamics. Imagine that we wish to study the dynamics of a large insect population in a cylindrical tube of length L and cross-sectional area A. (We could equally easily focus on bacteria, cells, or other organisms, and regard the tube as a canal or other one-dimensional medium.) From elementary differential equations we recall that such populations are often modeled by a growth law, or differential equation, for the population $u = u(t)$ of insects in the tube at time t. As observed in Section 1.1, the Malthus model is

$$\frac{du(t)}{dt} = ru(t),$$

which states that the growth rate is proportional the current population, where the proportionality constant r is the intrinsic growth rate. It is easy to solve this simple model equation using separation of variables and obtain a formula for the solution: $u(t) = u(0)e^{rt}$. Therefore, the model leads to exponentially increasing population growth. This model is reasonable for early stages of population growth, but it is clearly impossible for a population to continue in this growth mode for long times because resources (food, space, and other environmental factors) are generally limited. When resources are limited and there is competition for those resources, the Malthus model is often replaced by the logistics model

$$\frac{du(t)}{dt} = ru(t)\left(1 - \frac{u(t)}{K}\right),$$

where K is the carrying capacity. As the population $u(t)$ grows and approaches K, the growth rate approaches zero and there is a limit to the growth. The solution to this model equation is easily obtained by separation of variables, and it graphs as the familiar sigmoid, or S-shaped, curve.

In these two elementary dynamical models the population depends only on one independent variable, the time t. Also, although we usually say u is total population, it is implicitly understood that it is a population density, or population per unit volume (in this case, the fixed volume AL of the tube).

Clearly these models ignore how the insects are distributed spatially in the tube. If we wish to consider this distribution, then the population u will depend not only on time t but also upon a spatial variable x denoting position in the tube, where $0 \leq x \leq L$; that is, $u = u(x, t)$. With this interpretation, $u(x, t)$ is a local population density measuring the number of insects in a small volume $A\Delta x$ lying between x and $x + \Delta x$. In other words, $u(x, t)A\Delta x$ is the number of insects in the small volume. Implicit in this model are two facts: there are enough insects in the tube so that defining a local density makes sense, and there are no variations in the density in the radial direction. To model population changes, we must ask how it evolves in both time and space; therefore, we expect a PDE model for $u(x, t)$. If we adopt a growth model, then we must include terms (say, containing spatial partial derivatives) that indicate how the insects move in the tube. For example, if the motion is due to diffusion, then the PDE model for logistics growth is, as may be expected,

$$u_t = Du_{xx} + ru\left(1 - \frac{u}{K}\right), \qquad u = u(x,t).$$

This equation is called *Fisher's equation*, and it is one of the fundamental nonlinear PDEs of biology.

Finally, we observe that knowledge of the population density $u(x,t)$ permits computation of the total population in the tube at time t via

$$\text{Total insect population at time } t = \int_0^L u(x,t)A\,dx.$$

The preceding discussion indicates how consideration of spatial structure leads to a PDE model. There are other types of structure besides spatial structure. If we are interested in the age of the insects, then the population density depends upon the age a and time t; hence, $u = u(a,t)$, where $u(a,t)da$ represents the number of individuals at time t between the ages a and $a + da$. An important exercise in life history theory is to develop models that predict the age structure $u(a,t)$ of a population at any time t if the initial age structure $u(a,0)$ is known and the birth and death rates are given. These problems are discussed in Chapter 5. Other structured models might include other tags or labels like size or weight; these are called physiologically structured models. A population model that includes both age a and length l leads to a PDE model for a function of three variables, $u = u(a,l,t)$. So the population density depends upon the organism's age, its length, and time. Other PDE models in the life sciences include models for the biochemical reactions and motion of chemical species in the blood, in cells, and in other media.

There are two fundamental models in biology to describe motion: *advection* and *diffusion*. We encountered these processes in Sections 1.2 and 1.3. Advection is the more common term in biology, whereas convection is the term often used in engineering and other physical sciences. In the life sciences, advection refers to the transport of particles, chemicals, animals, or whatever, via bulk motion of the supporting medium, e.g., wind, water, blood, and so on. Diffusion refers to the random motion of particles, chemicals, animals, or whatever, that cause these entities to disperse from high to low concentrations. We shall discuss the random nature of diffusion in latter paragraphs. Figure 1.5 compares time profiles of both diffusion and advection processes, and a combination of the two.

A model of advection follows from the conservation law derived in Section 1.2. To reiterate, the basic conservation law in one dimension relates the time derivative of the unknown density $u = u(x,t)$, the spatial derivative of the unknown flux $\phi = \phi(x,t)$, and the source (or sink) $f =$

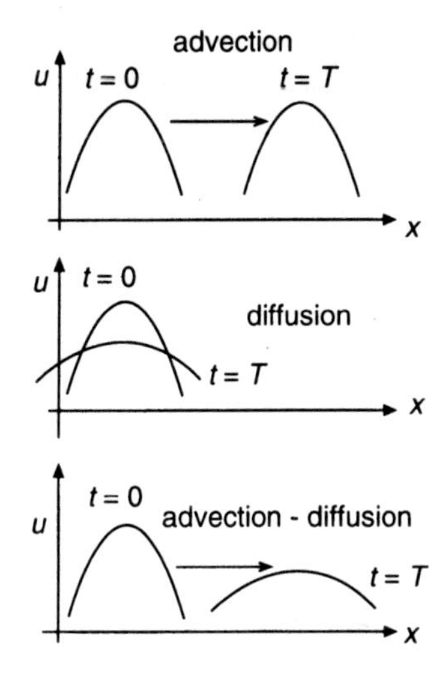

Figure 1.5. Spatial density profiles showing how an initial signal propagates in time under an advection, diffusion, or an advection–diffusion process.

$f(x, t)$. It is given by equation (1.8) as

$$u_t = -\phi_x + f. \tag{1.22}$$

The density can refer to density of animals, chemicals, or any particles or propagules. The flux, a measure of the flow rate, is usually dependent upon the density through the assumption of a constitutive equation; different motions arise from different assumptions. As we noted, *advection* is defined by the relation

$$\phi = cu,$$

where c is the bulk speed of the medium carrying the density, and *diffusion* is defined by Fick's law

$$\phi = -Du_x,$$

where D is the *diffusion coefficient*. The diffusion coefficient has dimensions of length-squared per time, and we often use the relation $D = L^2/T$ to estimate the time T that it takes for a substance to diffuse a length L. If c and D are constant, we can substitute these defining relations into (1.22), respectively, to obtain the fundamental balance equations:

$$u_t = -cu_x + f \qquad \text{(advection equation with source)}$$

$$u_t = Du_{xx} + f \qquad \text{(diffusion equation with source).}$$

If both advection and diffusion occur, then the flux is

$$\phi = -Du_x + cu,$$

and the fundamental balance law becomes

$$u_t = Du_{xx} - cu_x + f \qquad \text{(advection–diffusion equation with source).}$$

Sources and sinks depend upon the model. For population models the source term f represents birth or growth rates, and a sink represents a death rate, either by natural causes or by predation. For chemical problems, the source/sink term is called a reaction term and represents the rate that the chemical species is created or consumed by chemical reaction, or lost by absorption (say, across a cell boundary). Sources are positive terms, and sinks are negative terms. In addition, PDEs are always accompanied by a relevant domain of space and time and by initial conditions and/or boundary conditions; an initial condition prescribes the initial density distribution $u(x, 0)$, and a boundary condition prescribes what is going on at the boundary of the domain. In the latter case we may specify the density u itself, or we may specify the flux ϕ. It is important to notice that the flux is different for the three different models defined above: advection, diffusion, and advection–diffusion.

In some biological processes the advection speed c and the diffusion constant may depend upon location in the medium, and thus depend upon x, or they may depend upon the density u itself. That is, $c = c(x, u)$ and $D = D(x, u)$. Dependence upon x signifies a heterogeneous medium of propagation. Therefore, the advection–diffusion equation with a source is

$$u_t = (Du_x)_x - (cu)_x + f,$$

and the c and D factors may not be pulled out of the spatial derivatives. When c or D depends upon u, the equation is nonlinear.

The notion of diffusion was introduced in Section 1.3, mostly in the context of heat conduction. Now we discuss the connection between diffusion and random motion in biological settings.

In chemistry and physics it is easy to deduce, reasoning at an atomic or molecular level, how substances diffuse because of random motion. This microscopic description of diffusion is based on statistics and the fact that atoms or molecules collide randomly. These random collisions cause an assemblage of molecules to move from regions of high concentrations to regions of lower concentrations. For example, if a noxious gas is released at a point in a room, the colliding molecules of gas and the air lead to the eventual dispersion of the gas throughout the room. This stochastic, molecular model is the basis of the kinetic theory of gases, and it has a firm root in mathematics and in empirics.

In chemically based biological systems where we are tracking molecular concentrations, we observe the same phenomenon. For example, a chemical, toxic to fish, might be dumped locally in a reservoir; the spread of the chemical throughout the reservoir can be understood by the molecular collision model. In other biological systems, however, the same spreading effects are observed, but the cause is not random collisions on a microscopic scale, but interactions on a macroscopic scale. This effect is illustrated by the spread of an infectious disease. If infected individuals are introduced in a large population of susceptible individuals, then the infection can spread throughout the population by personal contact or by respiration of air borne particulates. Even in clumped populations of organisms, there is a tendency for individuals to migrate from the high-density regions to low-density regions; this macroscopic dispersion is the result of population 'pressures' like competition for resources or might, in humans, even have some psychological basis. But the observation is the same as in the molecular collision model—movement from high to low concentrations.

We modeled diffusion by *Fick's law* (or Fourier's law in heat conduction), which states that the flux, or motion, is proportional to the steepness of the concentration curve, i.e., $\phi(x,t) = -Du_x(x,t)$. So the migration is "down the concentration gradient," away from the most concentrated regions, and the greater the changes in concentration, the faster the motion. When this relation is substituted into the basic conservation law $u_t = -\phi_x$, we obtain the diffusion equation $u_t = Du_{xx}$.

Now we want to show how this diffusion model arises from a stochastic argument based upon random motion. Let X be a random variable with a normal probability density function (the "bell-shaped curve")

$$N(x,\mu,\sigma) = \frac{1}{\sqrt{2\pi\sigma^2}} e^{-(x-\mu)^2/2\sigma^2},$$

where μ is the mean, or center of the density, and σ is the standard deviation, a measure of the spread about the mean. By definition, the area under the normal curve, $\int_a^b N(x,\mu,\sigma)dx$, is the probability of X taking a value between a and b. Amazingly enough, if we take the standard deviation to depend upon time via $\sigma = \sqrt{2Dt}$, so that the spread of the data gets larger with time, then we obtain the function

$$u(x,t) = \frac{1}{\sqrt{4\pi Dt}} e^{-(x-\mu)^2/4Dt},$$

which is a solution to the diffusion equation (compare Exercise 1, Section 1.1)! This formula gives the solution to the diffusion equation generated by a unit point source given initially at $x = \mu$, and it is called the *fundamental solution* to the diffusion equation, or the *point-source solution*. That is, the point-source solution is the formula for the time-changing, spreading density distribution of a substance caused by placing a unit amount of the substance at $x = \mu$ at time $t = 0$.

It is not an accident that there is such a close relationship between the diffusion equation and probabilities. Let us look in a different direction to affirm this relationship.

Let x represent an arbitrary point on the x-axis. Now divide the axis into equal segments of length h so that the entire axis is composed of segments with discrete endpoints ..., $x-2h, x-h, x, x+h, x+2h$, ... Let $u = u(x,t)h$ be the number of particles (e.g., cells, organisms) in the interval $(x, x+h)$

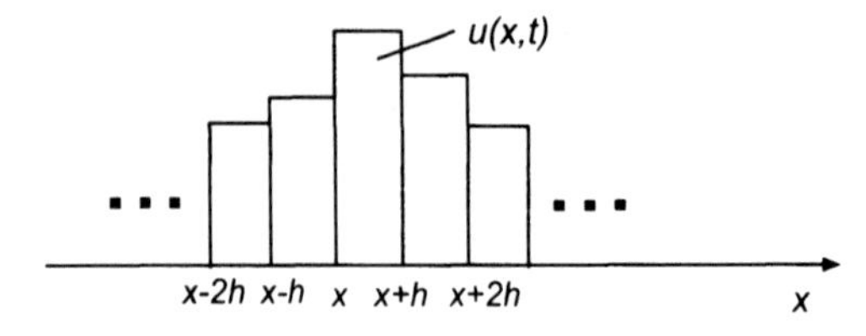

Figure 1.6. Histogram where $u(x, t)h$ represents the number of particles in the interval $(x, x + h)$ at time t.

at time t. Figure 1.6 is a schematic of a typical histogram showing the distribution of the assemblage of particles at time t. Now assume in the next small instant of time τ that all the particles in each interval move randomly to the right or to the left with an equal probability of one-half. Then we can calculate the particle distribution at the next time $t + \tau$. We get

Change in number of particles in $(x, x + h)$

= Number that move in from the left

+ Number that move in from the right

− Number that leave to the left

− Number that leave to the right.

Here, by "number" we mean "average number." In symbols, after canceling the common term h,

$$u(x, t + \tau) - u(x, t) = \frac{1}{2}u(x - h, t) + \frac{1}{2}u(x + h, t) - \frac{1}{2}u(x, t) - \frac{1}{2}u(x, t).$$

This simplifies to

$$u(x, t + \tau) = \frac{1}{2}u(x - h, t) + \frac{1}{2}u(x + h, t).$$

This is a difference equation relating the number of particles in three adjacent intervals at two different times. Because h and τ are small, we may expand the three terms of this equation in Taylor series to obtain

$$\begin{aligned} &u(x, t) + u_t(x, t)\tau + O(\tau^2) \\ &\quad = \frac{1}{2}\left(u(x, t) - u_x(x, t)h + \frac{1}{2}u_{xx}(x, t)h^2 + O(h^3)\right) \\ &\quad\quad + \frac{1}{2}\left(u(x, t) + u_x(x, t)h + \frac{1}{2}u_{xx}(x, t)h^2 + O(h^3)\right), \end{aligned}$$

where $O(\tau^2)$ denotes remaining terms of the series, all of which have a factor of at least τ^2, and $O(h^3)$ denotes remaining terms which have a factor of at least h^3. Simplifying yields

$$u_t(x, t) = \frac{h^2}{2\tau}u_{xx}(x, t) + O(\tau) + \frac{1}{\tau}O(h^3).$$

Now we take the limit as $\tau \to 0$ and $h \to 0$ in such a way that we maintain the ratio $\frac{h^2}{2\tau}$ equal to a fixed constant D. Therefore, in this special limit, we obtain

$$u_t(x, t) = Du_{xx}(x, t),$$

which is the diffusion equation. (Exercise 10 requests this same derivation when the probability of moving right or left is p and $1 - p$, respectively. One obtains the diffusion equation with "drift," i.e., the advection–diffusion equation.) Further discussion of the connection between diffusion and random motion can be found in Murray (2003).

Point-source solutions to linear diffusion equations play a special role in both applications and theory of PDEs, and so we make a few further comments and catalog some other such solutions. Point source solutions are also called *fundamental solutions*, and those that also satisfy given boundary conditions are called *Green's functions*. We shall show that knowledge of the point-source solution allows us to solve the initial value problem for the diffusion equation by superposition. If an advection or growth process is present, as well as diffusion, then we can also obtain special solutions that represent the density distribution arising from an initial point source. For the advection–diffusion equation

$$u_t = Du_{xx} - cu_x \tag{1.23}$$

in linear geometry, the point-source solution is

$$u = \frac{1}{\sqrt{4\pi Dt}} e^{-(x-\mu-ct)^2/4Dt} \tag{1.24}$$

while for the growth-diffusion equation

$$u_t = Du_{xx} + \gamma u, \tag{1.25}$$

the point-source solution is

$$u = \frac{1}{\sqrt{4\pi Dt}} e^{-(x-\mu)^2/4Dt+\gamma t}. \tag{1.26}$$

The diffusion equation in two dimensions with radial symmetry, where the density $u = u(r, t)$ depends only of the distance r from the origin and time, is (see Exercise 12)

$$u_t = D\frac{1}{r}(ru_r)_r.$$

The point-source solution is, with the unit source at the origin,

$$u(r, t) = \frac{1}{4\pi Dt} e^{-r^2/4Dt}.$$

In three dimensions, the diffusion equation with spherical symmetry, where $u = u(\rho, t)$ depends only upon the distance ρ from the origin and time, is

$$u_t = D\frac{1}{\rho^2}(\rho^2 u_\rho)_\rho.$$

In this case the point-source solution is, with the unit source at the origin,

$$u(\rho, t) = \frac{1}{(4\pi Dt)^{3/2}} e^{-\rho^2/4Dt}.$$

A major difference between point-source solutions in planar, radial, and spherical symmetry is the decaying, time-dependent amplitude factor appearing in front of the exponential term; the decay factors are $1/\sqrt{t}$, $1/t$, $1/t^{3/2}$, respectively. It is reasonable for signals to decay faster as the dimensions increase because there is more space for the material to go.

EXAMPLE

(*Invasion speed*) Consider a very long canal where a non-indigenous species of algae is accidentally released at $x = 0$. After a long period of time it is observed that the speed of the biological invasion outward in both directions from $x = 0$ is nearly constant. We show that this observation can be explained by a diffusion-growth model. The diffusion-growth model is given by (1.24) with point source solution given by (1.25). Let $x = x(t)$ denote the position of the wavefront, defined by the position where $u = u_f$, where u_f is a small, given value of the density. Then

$$\frac{1}{\sqrt{4\pi Dt}} e^{-x(t)^2/4Dt+\gamma t} = u_f$$

Taking logarithms,

$$\gamma t - \frac{x(t)^2}{4Dt} = \ln(\sqrt{4\pi Dt}u_f).$$

Now we make an interesting and clever approximation. For large times t the right side gets large, but not as large as the γt term (recall that t is much larger than $\ln t$). Therefore, the only two terms in this equation that can balance for large t are the two on the left side. Thus the long time, approximate position of the invasion front is

$$x(t) \approx \sqrt{4\gamma D}t.$$

Consequently, the speed of the front approaches a constant value $\sqrt{4\gamma D}$. □

Exercises

1. In Fick's law assume the diffusion coefficient D is a function of the density u, i.e., $D = D(u)$. In this case show that the diffusion equation without sources

can be written

$$u_t = D(u)u_{xx} + D'(u)u_x^2.$$

2. From the diffusion equation with no sources, find the steady-state concentration $u = u(x)$ of a chemical in a tube of length $L = 2$ with diffusion coefficient D and subject to boundary conditions $u(0) = 4$ and $u'(2) = 1$. Explain these conditions in terms of chemical concentrations and fluxes. Under the same boundary conditions, what is the steady-state concentration if D depends on x, e.g., $D = (1 + x)^{-1}$? What is the steady state solution if D depends on u, e.g., $D = u$?

3. Considering all cases, find the form of steady-state solutions to the advection–diffusion equation

$$u_t = Du_{xx} - cu_x$$

and the advection–diffusion–growth equation

$$u_t = Du_{xx} - cu_x + ru.$$

4. An environmentally toxic radioactive chemical is continually released at a constant rate of 1 (mg per vol per time) at the midpoint of a canal of length L. As it diffuses through the canal with diffusion constant $D = 1$, it decays at rate λ (per unit time) The ends of the canal are connected to large bodies of toxic-free water. Set up the model equations and the boundary conditions. Find the steady-state concentration and sketch its spatial profile for different values of L and λ. (Hint: Break the problem up into two parts, on each side of the source. At the source the concentration must be continuous, and in a small interval about the source, the "flux in" minus the "flux out" equals one).

5. Work the preceding problem if there is no decay and if the source is located at $x = \xi$, $0 < \xi < L$.

6. Consider the advection–diffusion equation on an interval $0 < x < L$. Show that if the flux at $x = 0$ equals the flux at $x = L$, then the density is constant.

7. The population density of zooplankton in a deep lake varies as a function of depth $x > 0$ and time t ($x = 0$ is the surface). Zooplankton diffuse vertically with diffusion constant D and buoyancy effects cause them to migrate toward the surface with an advection speed of αg, where g is the acceleration due to gravity. Ignore birth and death rates. Find a PDE model for the population density of zooplankton in the lake, along with the the appropriate boundary conditions at $x = 0$ and $x = +\infty$. Find the steady-state population density for zooplankton as a function of depth, and sketch its graph.

8. Sucrose in water has a diffusion coefficient of 4.6×10^{-6} cm^2/ sec. Estimate how far a concentrated amount of sucrose would diffuse in one day? A certain insect pheromone diffuses in air with diffusion constant 2.0×10^{-1} cm^2/ sec. Estimate the time it takes for the pheromone to diffuse a distance of 100 meters. Can a diffusion mechanism account for sexual attraction of mates for this insect?

9. Show that the steady-state solutions to the diffusion equation in linear geometry, in planar geometry with radial symmetry, and in spatial geometry with

spherical symmetry are, respectively,

$$u = ax + b,$$

$$u = a \ln r + b,$$

$$u = \frac{a}{\rho} + b.$$

Here, a and b are arbitrary constants.

10. Repeat the random walk derivation in this section in the case that the probability of moving right or left is p and $1 - p$, respectively, rather than one-half in either direction. Show that under appropriate limit definitions as $h \to 0$ and $\tau \to 0$, one obtains the advection–diffusion equation.

11. Muskrats were accidentally introduced in Europe in 1905 and the biological invasion spread approximately radially in all directions. Let u_f define a predetermined magnitude of the population density on a circle of radius $r = r(t)$ as the front spreads. Use a diffusion–growth model to show that the speed of the wave front is approximately constant for large times t.

12. Derive the radially symmetric diffusion equation

$$u_t = D \frac{1}{r} (ru_r)_r$$

from basic principles using the following strategy. Take two circles of radius $r = a$ and $r = b$ and write down the conservation law in polar coordinates for the density in the region in between (recall that an area element in polar coordinates is $r dr d\theta$ and note that the radial flux is $-Du_r$). Do the same for spherically symmetric diffusion

$$u_t = D \frac{1}{\rho^2} (\rho^2 u_\rho)_\rho$$

by considering two concentric spheres.

1.5 Vibrations and Acoustics

Wave motion is one of the most commonly occurring phenomena in nature. We mention electromagnetic waves, water waves, sound and acoustic waves, and stress waves in solids as typical examples. In this section we begin with a simple model to introduce the concept of wave motion and one of the fundamental PDEs, the wave equation, that describe such phenomena. The example is the small, transverse vibrations of a flexible, elastic string, e.g., a guitar string. Then we discuss wave motion in the context of acoustics.

Vibrations of a string

Let us imagine a taut string of length l fastened at its ends. Its motion is described by a function $u = u(x, t)$ that gives the vertical displacement

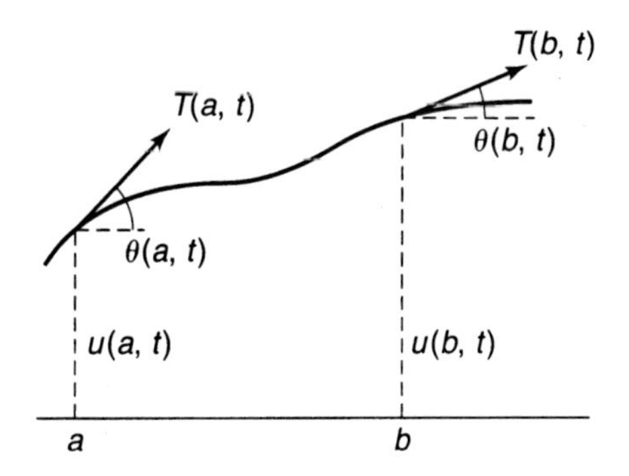

Figure 1.7. Displaced string with tension forces shown.

of each point x of the string at time t. Our basic postulate is that the displacement u is small. Implicitly, we assume that the motion is in a plane and no element of the string moves horizontally—only vertically. At each instant of time we assume that the string has a density $\rho(x,t)$, with dimensions mass per unit length, and the tension in the string is given by a function $T(x,t)$, with force dimensions. By convention, $T(x,t)$ is the force on the segment to the left of x caused by the portion of the string to the right of x, and we assume that the tension always is directed along the tangent to the profile at x. This latter assumption implies that the string does not resist bending. A priori, we do not know ρ or T. Figure 1.7 shows an arbitrary segment of the string between $x = a$ and $x = b$. We denote the angle that the tangent makes with the horizontal by $\theta(x,t)$, and we observe that

$$\tan\theta(x,t) = u_x(x,t).$$

To obtain an equation of motion of the string, we appeal to mass balance and Newton's second law. First, mass balance implies that the mass of the segment at any time t must equal its mass in equilibrium position (which we take to be at $t = 0$). In symbols, this means

$$\int_a^b \rho(x,t)\sqrt{1 + u_x(x,t)^2}dx = \int_a^b \rho_0(x)dx,$$

where ρ_0 is the density of the string in equilibrium. (Note that

$$\sqrt{1 + u_x(x,t)^2}\ dx$$

is an element of arc length.) But this equation must hold for every segment $[a,b]$, and therefore we may equate the integrands to obtain

$$\rho(x,t)\sqrt{1 + u_x(x,t)^2} = \rho_0(x). \tag{1.27}$$

Next let us apply Newton's second law, which states that the time rate of change of the total momentum of the segment must equal the net external force. We shall assume that the only force acting on the segment is tension caused by the rest of the string; we ignore gravity and any damping forces—see Exercises 1 and 2. Because there is no horizontal motion of the segment, the horizontal forces must balance, or

$$T(b,t)\cos\theta(b,t) - T(a,t)\cos\theta(a,t) = 0.$$

Because this equation must hold for all a and b, we must have

$$T(x,t)\cos\theta(x,t) = \tau(t)$$

for some function τ. Now, the time rate of change of the total momentum of the segment must equal the net vertical force. That is,

$$\frac{d}{dt}\int_a^b \rho(x,t)u_t(x,t)\sqrt{1+u_x(x,t)^2}dx$$
$$= T(b,t)\sin\theta(b,t) - T(a,t)\sin\theta(a,t).$$

Using the result of equation (1.27) and bringing the derivative inside the integral sign gives

$$\int_a^b \rho_0(x)u_{tt}(x,t)dx = T(b,t)\sin\theta(b,t) - T(a,t)\sin\theta(a,t).$$

We observe that the right side of this equation can be written as

$$T(b,t)\cos\theta(b,t)[\tan\theta(b,t) - \tan\theta(a,t)],$$

or

$$\tau(t)(u_x(b,t) - u_x(a,t)).$$

Therefore,

$$\int_a^b \rho_0(x)u_{tt}(x,t)dx = \tau(t)(u_x(b,t) - u_x(a,t)).$$

By the fundamental theorem of calculus we can rewrite the right side as an integral, and we obtain

$$\int_a^b \rho_0(x)u_{tt}(x,t)dx = \tau(t)\int_a^b u_{xx}(x,t)dx.$$

Again using the arbitrariness of the interval of integration (what we have said holds for any segment $[a,b]$), we can deduce

$$\rho_0(x)u_{tt}(x,t) = \tau(t)u_{xx}(x,t), \quad 0 < x < l, \ t > 0.$$

As an additional constitutive assumption, we assume that the tension $\tau(t) = \tau_0 =$ constant, which is a good assumption for small displacements of the string. Thus, in the framework of the assumptions that we have made, we obtain the final model equation for the transverse displacements, namely,

$$\rho_0(x)u_{tt} = \tau_0 u_{xx}. \tag{1.28}$$

If we introduce the function

$$c_0(x) = \sqrt{\tau_0/\rho_0(x)},$$

then we can write (1.28) as

$$u_{tt} = c_0(x)^2 u_{xx}. \qquad (1.29)$$

One should observe that c_0 has dimensions of speed. It is called the *wave speed*, and we shall observe that it measures the actual speed that waves travel along the string. In many applications the density of the string is a constant, and hence c_0 is constant; in that case equation (1.29) becomes

$$u_{tt} = c_0^2 u_{xx}, \qquad (1.30)$$

which is called the *wave equation*; it is one of the fundamental equations in mathematics and applications. If c_0 is constant, it is easy to check (Exercise 3) that $u = F(x - c_0 t)$ and $u = G(x + c_0 t)$ are solutions of the wave equation for any twice continuously differentiable functions F and G; thus the wave equation admits both right- and left-traveling wave solutions. In fact, we shall show later that the general solution of the wave equation is the superposition (sum) of left- and right-traveling waves.

Our original assumption that the string is fastened at its ends $x = 0$ and $x = l$ translates into *boundary conditions*

$$u(0, t) = 0, \quad u(l, t) = 0, \quad t \geq 0. \qquad (1.31)$$

To determine a unique displacement, we must impose initial conditions as well; that is, we prescribe both the position of the string and its velocity at time $t = 0$. Symbolically,

$$u(x, 0) = f(x), \quad u_t(x, 0) = g(x), \quad 0 \leq x \leq l. \qquad (1.32)$$

In summary, to determine the displacement $u(x, t)$, we must solve the initial boundary value problem consisting of the PDE (1.30) subject to the boundary conditions (1.31) and initial conditions (1.32).

We point out that this model is a small-displacement theory. A more complicated nonlinear model is required if we want to describe large displacements.

Acoustics

Acoustics is the science of sound and how signals are transmitted through a medium. We show that small disturbances, or sound signals, propagate as waves and are governed by the wave equation.

We restrict our attention to sound waves in a tube of cross-sectional area A, as in Figure 1.3, filled with a gas. In a fixed cross-section we assume no variation in any of the variables; the state variables will depend on x, the distance along the tube, and time t. Thus our treatment is strictly one-dimensional. The state variables are mass density $\rho = \rho(x, t)$, velocity

$v = v(x,t)$, and pressure $p = p(x,t)$. Here, the velocity is the actual velocity of the air particles as measured by a laboratory observer; the pressure is the force per unit area on the air to the right of the section at x, caused by the air to the left of the section at x.

The starting place for obtaining equations for acoustics is the basic balance law

$$u_t + \phi_x = f$$

introduced in Section 1.2. Here u is a density (quantity per unit volume), ϕ is the flux (quantity per unit area per unit time), and f is a source (quantity per unit volume per unit time). First, in gas flow we must balance mass. Therefore, take $u = \rho$, the mass density, and $\phi = \rho v$, which is the mass flux. There are no sources that create or destroy mass, so $f = 0$. Then *mass balance* is expressed by the equation

$$\rho_t + (\rho v)_x = 0. \tag{1.33}$$

Second, momentum must balance. Therefore, take $u = \rho v$, which is the momentum density (the same as the mass flux, in units of mass per area per time, or mass times velocity per unit volume), and take $\phi = (\rho v)v$, which is the momentum flux, or the rate at which momentum flows through a cross-section. But now there is a source term f because pressure gradients across the tube can create a net force that can change momentum (this is Newton's second law). So we take $f = -p_x$, and the *momentum balance law* is

$$(\rho v)_t + (\rho v^2)_x = -p_x.$$

Observe that the right side is a force per unit volume, which is the same as a momentum per unit volume per unit time; so it is like a momentum source term. The term is negative because a negative pressure gradient should cause motion to the right. We leave it as an exercise (Exercise 7) to show that the momentum balance equation can be written, using (1.33), as

$$\rho v_t + \rho v v_x + p_x = 0. \tag{1.34}$$

Equations (1.33) and (1.34) are the governing equations for gas flow in a tube. The equations are *nonlinear* and contain three unknowns, ρ, v, and p. For a third equation we assume a constitutive relation, or equation of state, of the form

$$p = F(\rho),$$

i.e., the pressure is a function of density. We will assume that $F'(\rho) > 0$, or pressure increases with density. A typical assumption is $p = k\rho^\gamma$, where $k > 0, \gamma > 1$. (If the equation of state should also depend on temperature, as in an ideal gas, then we would require yet another equation; this equation would be energy balance.)

Much of acoustical science deals with small disturbances in the gas. This leads to a simplified, linear approximation of equations (1.33)–(1.34). Let us assume that the gas in the tube is at rest, in a constant ambient state $\rho = \rho_0, v = 0, p_0 = F(\rho_0)$. Then we make a small disturbance, say at one end, and let $\rho = \rho_0 + \tilde{\rho}(x,t)$ and $v = \tilde{v}(x,t)$ be the resulting density and velocity, where $\tilde{\rho}$ and $\tilde{v}$, the deviations from the ambient state, are small. For the pressure we have, by a Taylor expansion,

$$p = F(\rho_0 + \tilde{\rho}) = F(\rho_0) + F'(\rho)\tilde{\rho} + \cdots = p_0 + c^2\tilde{\rho} + \cdots,$$

where

$$c \equiv \sqrt{F'(\rho_0)}$$

is called the *sound speed*, and the three dots denote terms containing at least a $\tilde{\rho}^2$ factor. Note that c has velocity units, since it is the square root of a derivative of pressure with respect to density. Substituting all these expressions into the mass and momentum balance laws (1.33)–(1.34) yields

$$(\rho_0 + \tilde{\rho})_t + \left((\rho_0 + \tilde{\rho})\tilde{v}\right)_x = 0,$$

$$(\rho_0 + \tilde{\rho})\tilde{v}_t + (\rho_0 + \tilde{\rho})\tilde{v}\tilde{v}_x + (p_0 + c^2\tilde{\rho} + \cdots)_x = 0.$$

Now we *linearize* these two equations by discarding all products of small terms; the idea is that products of small terms are smaller than the small terms themselves. Therefore, we obtain the approximation

$$\tilde{\rho}_t + \rho_0\tilde{v}_x = 0, \quad \rho_0\tilde{v}_t + c^2\tilde{\rho}_x = 0. \tag{1.35}$$

These linearized equations are called the *acoustic approximation* equations. To repeat, $\tilde{\rho}$ and $\tilde{v}$ are the small deviations from an ambient state ρ_0 and $v = 0$.

Now take the t-derivative of the first equation in (1.35) and the x-derivative of the second equation; upon subtracting the results we obtain

$$\tilde{\rho}_{tt} - c^2\tilde{\rho}_{xx} = 0.$$

This is the wave equation for the small deviation $\tilde{\rho}$. Similarly,

$$\tilde{v}_{tt} - c^2\tilde{v}_{xx} = 0.$$

In other words, small disturbances in the gas are propagated by the wave equation. We recall from Section 1.5 that the wave equation admits solutions of the form $F(x - ct)$ and $G(x + ct)$, or right- and left-traveling waves moving at speed c. For example, a small density or velocity disturbance can move to the right, undistorted, at speed c. Large-amplitude disturbances, for example caused by an explosion, do not propagate in this way, and the full nonlinear theory embodied in (1.33)–(1.34) must be applied. In this case waves can distort, and discontinuous solutions, called shock waves, can be produced.

Exercises

1. In the derivation of the wave equation (1.30) we assumed no gravitational force on the string. What would the model equation be if gravity were included? Note that gravity acts at each point of the string and derive the equation

$$u_{tt} = c_0^2 u_{xx} - g,$$

where g is the constant acceleration due to gravity.

2. Repeat the derivation in this section when, in addition, the vertical motion is retarded by a damping force proportional to the velocity of the string. Obtain the *damped wave equation*

$$u_{tt} = c_0(x)^2 u_{xx} - k u_t,$$

where k is the constant damping coefficient.

3. Let c_0 be constant. Verify that $u = F(x - c_0 t)$ and $u = G(x + c_0 t)$ are solutions of (1.30) for any twice-differentiable functions F and G. If $F(x) = 1/(1 + x^2)$, sketch time profiles of the wave

$$u(x, t) = \frac{1}{2}(F(x - t) + F(x + t)).$$

4. Consider the displacements of a string governed by the wave equation

$$u_{tt} = c^2 u_{xx}, \quad 0 \le x \le l,$$

where c is constant, and subject to the boundary conditions

$$u(0, t) = 0, \qquad u(l, t) = 0, \quad t \ge 0.$$

Show that for any positive integer n there is a solution of the form

$$u_n(x, t) = \cos\frac{n\pi ct}{l} \sin\frac{n\pi x}{l}.$$

What are the initial conditions? Describe the motion of the string in the cases $n = 1$ and $n = 2$ (sketch several fixed time snapshots of the string profile). In general, for any n, what is the temporal frequency of the oscillations represented by these solutions? These frequencies are called the fundamental frequencies, and the corresponding displacements are called the fundamental modes. How do the frequencies change as the length l of the string is changed? How do the frequencies depend on the tension? Discuss your answers in the context of a vibrating guitar string. These special solutions are called *standing waves*.

5. The total energy of the string governed by equation (1.28) with boundary conditions (1.31) is defined by

$$E(t) = \int_0^l \left(\frac{1}{2}\rho_0 u_t^2 + \frac{1}{2}\tau_0 u_x^2\right)dx.$$

Show that the total energy is constant for all $t \geq 0$. Hint: Multiply (1.28) by u_t and note that $(u_t^2)_t = 2u_t u_{tt}$ and $(u_t u_x)_x = u_t u_{xx} + u_{tx} u_x$. Then show

$$\frac{d}{dt}\int_0^l \rho_0 u_t^2 dx = \tau_0 u_t u_x \big|_0^l - \frac{d}{dt}\int_0^l \tau_0 u_x^2 dx.$$

6. If the equation of state has the form $p = F(\rho) = k\rho^\gamma$, find an expression for the sound speed c. Using this equation of state with $\gamma = 1.4$, compute the speed that sound waves travel in air. Take the density of air to be 0.0013 gm per cubic cm.

7. Using the mass balance equation (1.33), show that the momentum balance equation $(\rho v)_t + (\rho v^2)_x = -p_x$ can be written in the form (1.34).

8. At the end ($x = 0$) of a long tube ($x \geq 0$) the density of air changes according to the formula $\tilde{\rho}(0, t) = 1 - \cos 2t$ for $t \geq 0$, and $\tilde{\rho}(0, t) = 0$ for $t < 0$. Find a solution to the wave equation in the domain $x > 0$, $-\infty < t < \infty$, in the form of a right-traveling wave that satisfies the given boundary condition. Take $c = 1$ and plot the solution surface.

1.6 Quantum Mechanics

One of the fundamental equations of mathematical physics is the Schrödinger equation, which is the basic equation of quantum mechanics. In the next few paragraphs we give a brief description of Schrödinger's equation in one dimension.

Let us consider a particle of mass m moving on the x-axis under the influence of a continuous, conservative force given by $F(x)$ that depends only on the position x. According to the canon of classical particle mechanics, the motion $x = x(t)$ of the particle is governed by the dynamical equation

$$m\frac{d^2x}{dt^2} = F(x), \tag{1.36}$$

which is Newton's second law of motion (mass times acceleration equals force). If the initial position $x(0)$ and initial velocity dx/dt at $t = 0$ are known, then one can in theory solve the ODE model (1.36) subject to the initial conditions and actually determine the state of the particle, that is, its position and velocity, for all times $t > 0$. In this sense, classical mechanics is a *deterministic* model—knowledge of the initial state determines the future history of the particle.

In the early 1900s it was found that this classical, deterministic model of motion fails on the atomic scale. Quantum mechanics, which is a probabilistic model, grew out of the atomic physics revolution. Quantum theory dictates that the particle has no definite position or velocity; rather, one postulates a statistical, or probabilistic, interpretation of the state of the particle in terms of a *wave function* $\Psi(x, t)$. The square of the wave function, $|\Psi|^2$, is a probability density; that is,

$$\int_a^b |\Psi(x,t)|^2 dx$$

is interpreted as the probability of the particle being in the interval $a \leq x \leq b$ at time t. Thus $\int_{-\infty}^{\infty} |\Psi(x,t)|^2 dx = 1$, since the particle is located somewhere on the x-axis. From elementary probability theory it is known that the probability density $|\Psi(x,t)|^2$ contains all of the statistical information for a given problem (for example, the mean and variance of the position). Note that Ψ is complex-valued, and thus $|\Psi|^2 = \Psi\overline{\Psi}$, where the overbar denotes complex conjugation.

So the question is how to find the wave function. The equation that governs the evolution of a quantum-mechanical system (the analogue of (1.36) for a classical system) is the Schrödinger equation, a second-order partial differential equation having the form

$$i\hbar\Psi_t = -\frac{\hbar^2}{2m}\Psi_{xx} + V(x)\Psi, \quad x \in R, \ t > 0, \tag{1.37}$$

where $V = V(x)$ is the potential energy, m is the mass, and $\hbar = h/(2\pi)$, where $h = 6.625 \cdot 10^{-34}$ kg m^2/s is Planck's constant. (Recall that associated with the force $F(x)$ is a potential function $V(x)$ defined by the equation $F(x) = -V'(x)$; that is, the force is the negative gradient of the potential.) One can motivate the Schrödinger equation from momentum and energy considerations, but here our goal is only to note the form of the equation. The reader is referred to one of the many excellent treatises on quantum mechanics for a complete discussion. A popular equation studied in detail in the mathematical literature is the *free* Schrödinger equation

$$\Psi_t = i\Psi_{xx}, \tag{1.38}$$

where the potential V is taken to be zero (a free particle with no forces acting); the constants are taken to be unity. Note that this equation resembles the diffusion equation, but it has a complex coefficient; this makes solutions of the two equations quite different.

One method used to find solutions of (1.37) (we shall observe later that this method is basic in PDEs) is to assume that the variables separate, i.e., the solution has the form of a product $\Psi(x,t) = y(x)\phi(t)$. Substituting this into (1.37) gives, after rearrangement,

$$\frac{i\hbar\phi'(t)}{\phi(t)} = \frac{-\frac{\hbar^2}{2m}y''(x) + V(x)y(x)}{y(x)}.$$

Here, prime denotes the derivative. The left side of this equation depends only on t, and the right side depends only on x. The only way equality can occur for all t and x is if both sides are equal to the same constant, which we shall call E (for "energy"). Therefore, we obtain two equations, one for ϕ,

$$\frac{d\phi}{dt} = (-iE/\hbar)\phi, \tag{1.39}$$

and one for y, namely,

$$-\frac{\hbar^2}{2m}y'' + (V(x) - E)y = 0. \tag{1.40}$$

The solution of (1.39) is easily found to be

$$\phi = C\exp(-iEt/\hbar) = C(\cos(-Et/\hbar) - i\sin(Et/\hbar)),$$

where C is a constant. Thus, the temporal part is oscillatory (periodic). Equation (1.40), whose solution $y(x)$ gives the spatial part of the wave function, is called the *time-independent Schrödinger equation*, and it is one of the fundamental ODE models of mathematical physics. The values of E for which (1.40) has a nontrivial solution $y(x)$ with $\int_{-\infty}^{\infty} y(x)^2 dx = 1$ are interpreted as the allowable energy levels. This is in contrast to the situation in classical mechanics, where any energy level is possible.

Exercises

1. Show that the ordinary differential equation $y'' + p(x)y' + q(x)y = 0$ can be transformed into the Schrödinger-like equation $u'' + r(x)u = 0$, $r = q - p'/2 - p^2/4$, without a first derivative, by the Liouville transformation $u = y\exp(\frac{1}{2}\int_a^x p(\xi)d\xi)$.

2. Show that $\int_{-\infty}^{\infty} |\Psi(x,t)|^2 dx = \int_{-\infty}^{\infty} y(x)^2 dx$.

3. If $V(x) = \frac{k}{2}x^2$, then for a special value of E, equation (1.40) admits a solution of the form $y(x) = Ce^{-ax^2}$ for some $a > 0$ and any constant C. Determine a and E, and sketch a graph of the probability density function $|\Psi|^2$ in the case $k = m = \hbar$ (use a calculator to compute C). Numerically calculate the probability that the particle is in the interval $[0, 1]$.

4. Suppose $\Psi(x,t)$, $x \in \mathbf{R}$, $t > 0$, is a twice continuously differentiable function that satisfies (1.35), and both $|\Psi|$, $|\Psi_x| \to 0$ as $|x| \to \infty$. Show that

$$\int_{-\infty}^{\infty} |\Psi(x,t)|^2 dx = \text{constant}, \quad t > 0.$$

Hint: Take the time derivative of the left side, pulling it under the integral; use the Schrödinger equation and its complex conjugate, and finally integrate the terms like $\overline{\Psi}_{xx}\Psi$ by parts.

5. A free particle of mass m is confined to the interval $0 < x < \pi$, and $\Psi(0, t) = \Psi(\pi, t) = 0$ for all t. Show that the associated time-independent problem is

$$y'' + \frac{2mE}{\hbar^2} y = 0, \quad 0 < x < \pi; \quad y(0) = y(\pi) = 0.$$

Show that this problem has a nontrivial solution if and only if E is one of the values $n^2\hbar^2/(2m)$, $n = 1, 2, \ldots$. Sketch the probability density functions $y(x)^2$ corresponding to $n = 1$ and $n = 2$, and compute the probability that the particle is in the interval $[0, 0.25]$ in each case.

1.7 Heat Flow in Three Dimensions

So far, our PDE models have involved one spatial dimension and time. It should be clear to the reader that many, if not most, physical problems occur in more than one spatial dimension. In this section we set up a PDE model for heat flow in three dimensions. The same ideas carry over to diffusion of chemicals or other biological quantities. This section may require the reader to review some of the concepts from multi-variable calculus.

The idea is exactly the same as in one dimension, namely to formulate an integral conservation law. Let Ω be a region in space where heat is flowing, and let $u = u(x, y, z, t)$ be the temperature at time t at the point (x, y, z) in Ω. We assume that the region is homogeneous and is characterized by a constant specific heat c and a constant density ρ. Now let B be an arbitrary sphere contained in Ω. See Figure 1.8.

We shall apply an energy balance principle to B that requires that the rate of change of the total amount of heat energy in B must equal the rate at which heat flows into B across its boundary plus the rate at which heat energy is generated by any sources in B. The total amount of heat in a small volume element $dV = dxdydz$ is $c\rho u dV$, and thus the total amount

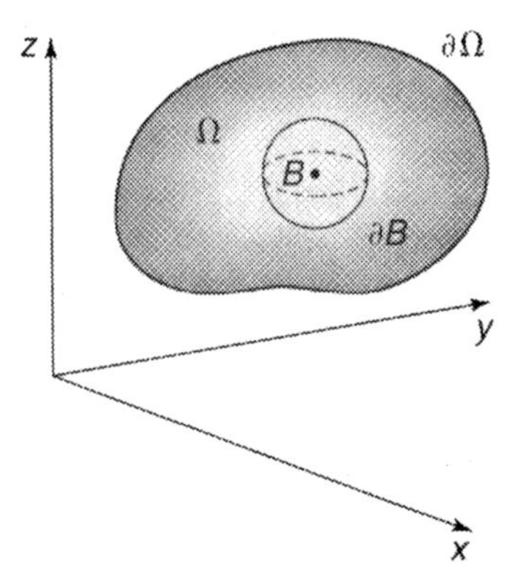

Figure 1.8. Arbitrary ball, or sphere, B contained in a region Ω. ∂B denotes the boundary of B.

of heat energy in B is given by the three-dimensional integral

$$\int_B c\rho u \, dV.$$

This integral is sometimes written as a triple integral $\int\int\int_B$, but for conciseness we shall adopt the single-integral notation. We assume that the heat sources (or sinks) are given by a point function $f = f(x, y, z, t)$, where $f dV$ is the rate at which heat is generated in dV; thus, the rate at which heat is generated in the whole of B is

$$\int_B f \, dV.$$

Note that f has dimensions of energy per unit volume, per unit time.

Next, we introduce the heat flux vector $\phi = \phi(x, y, z, t)$; its direction corresponds to the direction of heat flow at position (x, y, z) at time t. The rate at which heat flows across an oblique surface element dA oriented by a unit vector n is

$$\phi \cdot n \, dA.$$

See Figure 1.8.

Consequently, the net rate that heat energy flows across the boundary of B, denoted by ∂B, is the surface integral

$$\int_{\partial B} \phi \cdot n \, dA.$$

So, the conservation law, or energy balance law, is

$$\frac{d}{dt} \int_B c\rho u \, dV = -\int_{\partial B} \phi \cdot n \, dA + \int_B f \, dV. \tag{1.41}$$

The minus sign appears in front of the flux integral on the right because if the flow is outward, the integral is positive; the minus sign will cause the net energy on the left side of the equation to decrease, which is correct. Now we use one of the fundamental integral relations from multi-variable calculus—the divergence theorem. It permits us to rewrite the flux integral in (1.41) as a volume integral. The *divergence theorem*, a three-dimensional version of the fundamental theorem of calculus, states that under sufficient differentiability requirements on the vector field ϕ,

$$\int_B \text{div}\phi \, dV = \int_{\partial B} \phi \cdot n \, dA.$$

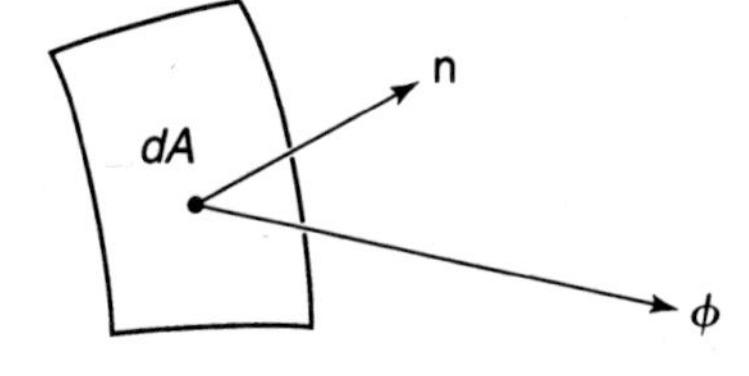

Figure 1.9. Heat flux through a surface element dA oriented by its unit normal vector n is $\phi \cdot n \, dA$.

Recall that if $\phi = (\phi_1, \phi_2, \phi_3)$, then its divergence is the scalar function

$$\text{div } \phi = \frac{\partial \phi_1}{\partial x} + \frac{\partial \phi_2}{\partial y} + \frac{\partial \phi_3}{\partial z}.$$

Thus we can write equation (1.41) as

$$\frac{d}{dt}\int_B c\rho u \, dV = -\int_B \text{div } \phi \, dV + \int_B f dV.$$

Now we can bring the time derivative under the integral on the left side and finally rearrange all the terms under one volume integral as

$$\int_B (c\rho u_t + \text{div } \phi - f) dV = 0.$$

This balance law must hold for every sphere B in Ω, and therefore the integrand must vanish, giving the partial differential equation

$$c\rho u_t + \text{div } \phi = f \tag{1.42}$$

for all t and all $(x, y, z) \in \Omega$. Equation (1.42) is the local (as opposed to the integral) form of the conservation law. It still contains two unknowns, the scalar temperature u and the vector heat flux ϕ. A constitutive relation can be postulated to connect the two. One such relation is the three-dimensional version of Fick's law, or Fourier's heat conduction law, which states, consistent with the laws of thermodynamics, that heat flows down the gradient, or in symbols,

$$\phi = -K \text{ grad } u = -K(u_x, u_y, u_z). \tag{1.43}$$

Recall from calculus that the negative gradient of a function points in the direction of maximum decrease. The proportionality constant K is the thermal conductivity. Substituting (1.43) into (1.42) and using the identity

$$\text{div grad } u = u_{xx} + u_{yy} + u_{zz},$$

we obtain the heat equation for the temperature $u = u(x, y, x, t)$ in three dimensions:

$$c\rho u_t - K(u_{xx} + u_{yy} + u_{zz}) = f. \tag{1.44}$$

The expression $u_{xx} + u_{yy} + u_{zz}$ is called the *Laplacian* of u, and it is denoted by Δu. That is,

$$\Delta u = u_{xx} + u_{yy} + u_{zz}.$$

In some books the Laplacian of u is denoted by $\nabla^2 u$. Finally, in summary, we can write (1.41) as

$$u_t - k\Delta u = \frac{1}{c\rho} f, \tag{1.45}$$

where the constant $k = K/(c\rho)$ is called the diffusivity.

Of course, these calculations can be made in two dimensions in order to describe heat flow in the plane. The two-dimensional analogue of (1.45) is the same,

$$u_t - k\Delta u = \frac{1}{c\rho} f, \tag{1.43}$$

where now $u = u(x, y, t)$, $f = f(x, y, t)$, and Δu is the *two-dimensional Laplacian*

$$\Delta u = u_{xx} + u_{yy}.$$

Generally, we expect that the three-dimensional heat equation (1.45) will come with auxiliary conditions in the form of an initial temperature condition

$$u(x, y, z, 0) = u_0(x, y, z) \quad \text{in } \Omega$$

and conditions on the boundary $\partial\Omega$.

If the source term f and the boundary conditions on $\partial\Omega$ are independent of time, then we would expect that the body Ω would eventually come to an equilibrium temperature $u = u(x, y, z)$ that is independent of time. This would occur long after the transients, or effects of the initial condition, decay away. From (1.42), since $u_t = 0$, we infer that this equilibrium temperature will satisfy the PDE

$$-k\Delta u = \frac{1}{c\rho} f,$$

which is *Poisson's equation*. If there are no sources, $f(x, y, z) \equiv 0$, then $u = u(x, y, z)$ satisfies *Laplace's equation*

$$\Delta u = 0 \quad \text{in } \Omega. \tag{1.47}$$

Laplace's equation, which is discussed in the next section, is one of the most famous models in the mathematical sciences, and it governs many phenomena besides equilibrium temperatures.

Because Laplace's equation (1.47) does not contain time, we expect that only boundary conditions should be appended in order to have a physically well-posed problem. We can prescribe the temperature on the boundary of Ω, that is,

$$u(x, y, z) = g(x, y, z), \quad (x, y, z) \in \partial\Omega,$$

which is called a Dirichlet condition; or we can prescribe the heat flux on the boundary, that is,

$$-K \operatorname{grad} u \cdot n = h(x, y, z), \quad (x, y, z) \in \partial\Omega,$$

which is a Neumann condition. If no heat flows across the boundary, then $h = 0$, and we say that the boundary is insulated.

Exercises

1. Let $u = u(x, y, z)$ be a scalar field. Verify that

$$\text{div}\,(\text{grad}\, u) = u_{xx} + u_{yy} + u_{zz}.$$

2. What is the form of the conservation law (1.42) if the thermal conductivity K in Fourier's law (1.43) depends on position, i.e., $K = K(x, y, z)$? (In this case we say that the medium is heterogeneous.)

3. Suppose $u = u(x, y, z)$ is a solution of the Neumann boundary value problem

$$-K\Delta u = f, \quad (x, y, z) \in \Omega,$$

$$-K\text{grad}\, u \cdot n = g(x, y, z), \quad (x, y, z) \in \partial\Omega,$$

where f and g are functions of x, y, z. Show that f and g must satisfy the relation

$$\int_{\Omega} f dV = \int_{\partial\Omega} g dA.$$

In terms of steady heat flow, what is the physical meaning of this relation?

4. Let w be a scalar field and ϕ a vector field. Prove the vector identity

$$\text{div}\,(w\phi) = \phi \cdot \text{grad}\, w + w\, \text{div}\, \phi.$$

Integrate this equation over Ω and take $\phi = \text{grad}\, u$, where u is a scalar field, to prove *Green's identity*

$$\int_{\Omega} w\Delta u\, dV = -\int_{\Omega} \text{grad}\, u \cdot\, \text{grad}\, w\, dV + \int_{\partial\Omega} w\, \text{grad}\, u \cdot n\, dA.$$

5. Prove that if the Dirichlet problem

$$\Delta u = \lambda u, \quad (x, y, z) \in \Omega,$$

$$u = 0, \quad (x, y, z) \in \partial\Omega,$$

has a nontrivial solution $u = u(x, y, z)$, then λ must be negative. Hint: Multiply the equation by u and integrate over Ω; use Green's identity in Exercise 4.

6. (Dirichlet's principle) Suppose that u satisfies the Dirichlet problem

$$\Delta u = 0, \quad (x, y, z) \in \Omega,$$

$$u = f, \quad (x, y, z) \in \partial\Omega.$$

Show that ·

$$\int_{\Omega} |\, \text{grad}\, u\, |^2\, dV \le \int_{\Omega} |\, \text{grad}\, w\, |^2\, dV$$

for any other function w that satisfies $w = f$ on $\partial\Omega$. Thus, the solution to the Dirichlet problem minimizes the "energy"

$$\int_{\Omega} |\, \text{grad}\, w\, |^2\, dV.$$

Hint: Let $w = u + v$, where $v = 0$ on $\partial\Omega$, and expand $\int_{\Omega} |\, \text{grad}\, w\, |^2\, dV$ using Green's identity.

7. If heat is both diffusing and advecting in a spatial domain with flux given by $\phi = -K\text{grad}u + u \cdot v$, where v is nonconstant vector advection velocity, show that the governing advection–diffusion equation is given by

$$c\rho u_t = K\Delta u - (\text{div}v)u - v \cdot \text{grad}u.$$

1.8 Laplace's Equation

In the last section we derived Laplace's equation:

$$\Delta u = u_{xx} + u_{yy} = 0 \quad \text{in two dimensions;}$$

$$\Delta u = u_{xx} + u_{yy} + u_{zz} = 0 \quad \text{in three dimensions.}$$

We observed that Laplace's equation models steady heat flow in a region where the temperature is fixed on the boundary. Now, working in two dimensions, we show what Laplace's equation tells us physically. Let (x, y) be some point in the region where heat is flowing and let h be some small distance. By Taylor's theorem we have

$$u(x-h, y) = u(x, y) - hu_x(x, y) + \frac{1}{2}h^2 u_{xx}(x, y) + O(h^3),$$

where $O(h^3)$ denotes the remaining terms (the error term), which are at least cubic in h. Moreover,

$$u(x+h, y) = u(x, y) + hu_x(x, y) + \frac{1}{2}h^2 u_{xx}(x, y) + O(h^3).$$

If we add these two expressions and solve for u_{xx}, we obtain

$$u_{xx}(x, y) = \frac{u(x-h, y) - 2u(x, y) + u(x+h, y)}{h^2} + O(h^2),$$

where $O(h^2)$ denotes terms that are quadratic in h. This expression on the right side is a difference-quotient approximation to the second partial derivative u_{xx} at (x, y), and the error is proportional to h^2. Performing the same calculation, but now incrementing y instead of x, we obtain the approximation

$$u_{yy}(x, y) = \frac{u(x, y-h) - 2u(x, y) + u(x, y+h)}{h^2} + O(h^2).$$

Therefore, Laplace's equation at (x, y) can be approximated by the equation

$$\begin{aligned}\Delta u(x, y) = {} & \frac{u(x-h, y) - 2u(x, y) + u(x+h, y)}{h^2} \\ & + \frac{u(x, y-h) - 2u(x, y) + u(x, y+h)}{h^2} + O(h^2) \\ = {} & 0.\end{aligned}$$

Solving for $u(x, y)$ gives, upon neglecting the small-order terms,

$$u(x, y) \approx \frac{1}{4}(u(x-h, y) + u(x+h, y) + u(x, y-h) + u(x, y+h)). \quad (1.48)$$

Equation (1.48) states, physically, that the temperature at a point (x, y) is approximately (we neglected small terms) the average of the temperatures at the four nearby points $(x-h, y)$, $(x+h, y)$, $(x, y-h)$, $(x, y+h)$. See Figure 1.10. In Exercise 6 and in Chapter 4 we show how this formula can be used to approximate steady temperatures. Also, we observe that the temperature at (x, y) cannot exceed the temperatures at the neighboring points; so a maximum could not occur at (x, y). Similarly, a minimum could not occur at (x, y).]

This important physical interpretation can be extended to a circle. If $u = u(x, y)$ satisfies Laplace's equation at a point $P_0 : (x_0, y_0)$ in a region, then the temperature at P_0 is the average value of the temperature on any circle $C_R(P_0)$ of radius R centered at P_0 lying in the region; in symbols,

$$u(x_0, y_0) = \frac{1}{2\pi R}\int_{C_R(P_0)} u(x, y)ds.$$

The integral here is a line integral over the curve $C_R(P_0)$. (Of course, this result generalizes to three dimensions: The steady-state temperature at a point is the average of the temperatures over the surface of any sphere centered at that point.)

If the value of u at a point is the average of the values on any circle about that point, then the value of u at that point cannot exceed every value of u on any given circle. Intuitively, this seems to imply that a function u satisfying Laplace's equation in a given domain cannot have a local maximum at a point inside that domain; the maximum must therefore occur on the boundary of the domain. Indeed, this is true, and the result is called the maximum principle. We state the result precisely.

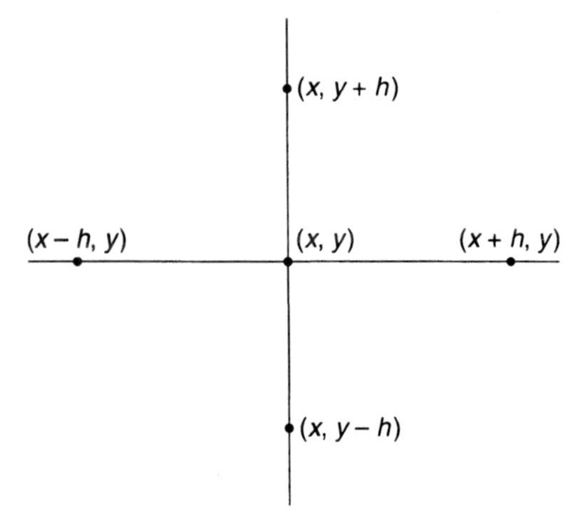

Figure 1.10. If u satisfies Laplace's equation, then the temperature at (x, y) is approximately the average of the temperatures at the four nearby points.

Theorem (The maximum principle)
Let $u(x,y)$ satisfy Laplace's equation in D, an open, bounded, connnected region in the plane; and let u be continuous on the closed domain $D \cup \partial D$ consisting of D and its boundary. If u is not a constant function, then the maximum and minimum values of u are attained on the boundary of D and nowhere inside D.

This result is also true in three dimensions. It is one of the fundamental properties of solutions to Laplace's equation.

Many two-dimensional problems involving Laplace's equation are in regions (circles, wedges, annuli) that lend themselves to a polar description in terms of r and θ, rather than rectangular coordinates x and y. This means that we need an expression for the Laplacian in terms of polar coordinates; the connection between polar and rectangular coordinates is given by the familiar formulae

$$x = r\cos\theta, \quad y = r\sin\theta,$$

or, inversely,

$$r = \sqrt{x^2 + y^2}, \quad \theta = \arctan\frac{y}{x}.$$

The chain rule allows us to calculate the partial derivatives u_{xx} and u_{yy} in terms of derivatives with respect to r and θ. First, we calculate $\partial u/\partial x$; we have

$$\frac{\partial u}{\partial x} = u_r r_x + u_\theta \theta_x = (\cos\theta)u_r - \frac{\sin\theta}{r}u_\theta.$$

So, an x partial derivative produces the operation

$$\frac{\partial}{\partial x} = \cos\theta\frac{\partial}{\partial r} - \frac{\sin\theta}{r}\frac{\partial}{\partial\theta}$$

in polar coordinates. Therefore, the second partial is

$$\begin{aligned}\left(\frac{\partial}{\partial x}\right)^2 &= \left(\cos\theta\frac{\partial}{\partial r} - \frac{\sin\theta}{r}\frac{\partial}{\partial\theta}\right)^2 \\ &= \cos^2\theta\frac{\partial^2}{\partial r^2} - \frac{2\sin\theta\cos\theta}{r}\frac{\partial^2}{\partial r\partial\theta} \\ &\quad + \frac{\sin^2\theta}{r^2}\frac{\partial^2}{\partial\theta^2} + \frac{2\cos\theta\sin\theta}{r^2}\frac{\partial}{\partial\theta} + \frac{\sin^2\theta}{r}\frac{\partial}{\partial r}.\end{aligned}$$

The reader should verify these calculations. Similarly, one finds that

$$\left(\frac{\partial}{\partial y}\right)^2 = \left(\sin\theta\frac{\partial}{\partial r} + \frac{\cos\theta}{r}\frac{\partial}{\partial\theta}\right)^2$$

$$= \sin^2\theta\frac{\partial^2}{\partial r^2} + \frac{2\sin\theta\cos\theta}{r}\frac{\partial^2}{\partial r\partial\theta}$$

$$+ \frac{\cos^2\theta}{r^2}\frac{\partial^2}{\partial\theta^2} - \frac{2\cos\theta\sin\theta}{r^2}\frac{\partial}{\partial\theta} + \frac{\cos^2\theta}{r}\frac{\partial}{\partial r}.$$

Upon adding these two equations there is a lot of cancellation, and we obtain

$$\Delta u = u_{rr} + \frac{1}{r}u_r + \frac{1}{r^2}u_{\theta\theta} = 0,$$

which is *Laplace's equation in polar coordinates.*

In three dimensions, problems often occur on regions that have cylindrical symmetry (e.g., portions of cylinders) or spherical symmetry (e.g., portions of spheres). Therefore, it is useful to have expressions for the Laplacian in cylindrical coordinates r, θ, z and spherical coordinates ρ, θ, ϕ. Cylindrical coordinates in space are defined by the equations

$$x = r\cos\theta, \quad y = r\sin\theta, \quad z = z,$$

which are just polar coordinates in the plane appended by z. The Laplace's equation in cylindrical coordinates is

$$\Delta u = u_{rr} + \frac{1}{r}u_r + \frac{1}{r^2}u_{\theta\theta} + u_{zz} = 0.$$

Spherical coordinates (see Figure 1.11) are related to rectangular coordinates by the transformation formulae

$$x = \rho\sin\phi\cos\theta, \quad y = \rho\sin\phi\sin\theta, \quad z = \rho\cos\phi.$$

Laplace's equation in spherical coordinates is

$$\Delta u = u_{\rho\rho} + \frac{2}{\rho}u_\rho + \frac{1}{\rho^2\sin\phi}(\sin\phi u_\phi)_\phi + \frac{1}{\rho^2\sin^2\phi}u_{\theta\theta} = 0.$$

Clearly, it is a tedious calculation using the chain rule to derive this formula, and we leave its verification as an exercise.

When symmetries are involved, these expressions for the Laplacian can simplify significantly.

EXAMPLE

Find the steady temperatures in the region between two concentric spheres of radii $\rho = 1$ and $\rho = 2$ if the inner sphere is held at zero degrees and the outer sphere is held at 10 degrees. Because there is no variation of the temperature on the boundaries in the longitudinal (θ) direction or in the azimuthal (ϕ) direction, it is clear that the steady temperature depends

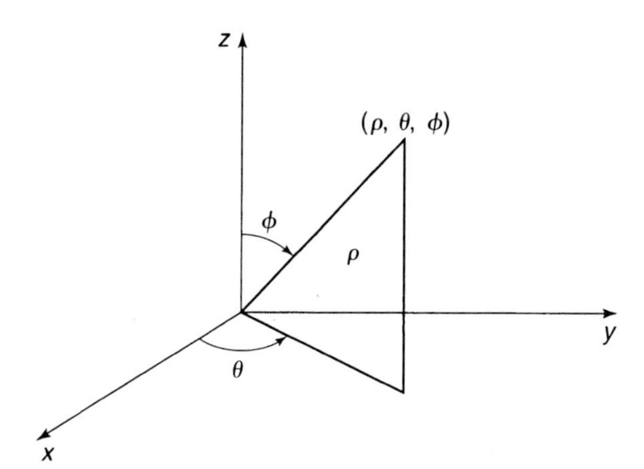

Figure 1.11. Spherical coordinates.

only on the distance ρ from the center of the coordinate system. That is, $u = u(\rho)$. Then, in spherical coordinates, Laplace's equation simplifies to

$$u''(\rho) + \frac{2}{\rho}u'(\rho) = 0.$$

This is the same as

$$\frac{d}{d\rho}(\rho^2 u') = 0.$$

Integrating twice with respect to ρ gives

$$u(\rho) = \frac{a}{\rho} + b,$$

where a and b are constants of integration. Now, $u(1) = 0$ forces $b = -a$, and then $u(2) = 10$ forces $a = -20$. Therefore, the steady temperatures vary radially according to

$$u(\rho) = 20\left(1 - \frac{1}{\rho}\right).$$

□

Exercises

1. In two dimensions suppose $u = u(r, \theta)$ satisfies Laplace's equation $\Delta u = 0$ in the disk $0 \le r < 2$, and on the boundary it satisfies $u(2, \theta) = 3\sin 2\theta + 1$, $0 \le \theta < 2\pi$. What is the value of u at the origin? Where do the maximum and minimum of u occur in the closed domain $0 \le r \le 2$?

2. Laplace's equation in one dimension is $u''(x) = 0$, which has general solution $u(x) = ax + b$, where a and b are constants. Find the steady-state temperature distribution in a metal bar of length l whose ends at $x = 0$ and $x = l$ are

held at temperatures T_0 and T_1, respectively. Sketch a graph showing this distribution. Suppose the right end at $x = l$ is insulated; now what is the steady-state temperature distribution if the left end is held at T_0 degrees?

3. Suppose a metal bar of unit length has a variable conductivity $K(x)$ given by $K(x) = 1 + x^2$, $0 \le x \le 1$. What is the steady-state temperature distribution $u(x)$ if $u(0) = 1$ and $u(1) = 4$? (Note that steady temperatures are modeled by the equation $-(K(x)u')' = 0$).

4. Find all radial solutions to the two-dimensional Laplace's equation. That is, find all solutions of the form $u = u(r)$, where $r = \sqrt{x^2 + y^2}$. Find the steady-state temperature distribution in the annular domain $1 \le r \le 2$ if the inner circle $r = 1$ is held at 0 degrees and the outer circle $r = 2$ is held at 10 degrees.

5. In a three-dimensional spherical region D free of charges, the static electric-field vector E satisfies the two Maxwell equations

$$\text{div } E = 0, \quad \text{curl } E = 0$$

(the electric field is produced, say, by electrical charges outside of D). Let $V = V(x, y, z)$ be the electrical potential defined by grad $V = E$ (review multi-variable calculus to see why such a potential exists). Show that $\Delta V = 0$. That is, the static electric field potential satisfies Laplace's equation.

6. This exercise illustrates an important numerical procedure for solving Laplace's equation on a rectangle. Consider Laplace's equation on the rectangle $D: 0 < x < 4,\ 0 < y < 3$ with boundary conditions given on the bottom and top by $u(x, 0) = 0,\ u(x, 3) = 0$ for $0 \le x \le 4$ and boundary conditions given on the sides by $u(0, y) = 2y(3 - y),\ u(4, y) = 0$ for $0 \le y \le 4$. Apply the average value property (1.45) with $h = 1$ at each of the six lattice points $(1, 1), (1, 2), (2, 1), (2, 2), (3, 1), (3, 2)$ inside D to obtain a system of six equations for the six unknown temperatures $u(i, j),\ i = 1, 2, 3;\ j = 1, 2$. Solve the system to approximate the steady temperature distribution and plot the resulting approximate surface using a software package.

1.9 Classification of PDEs

An ordinary differential equation is classified according to its order (the highest derivative that appears) and whether it is linear or nonlinear. PDE models are more difficult to classify. Not only are order and linearity important issues but the PDE structure is equally important as well. In fact, the structure of a PDE dictates what types of boundary and initial conditions can be imposed and ultimately what types of physical processes the PDE models.

What do we mean by the structure of a PDE? To fix the notion, let us consider second-order equations. We have already obtained three model equations:

$$u_t - ku_{xx} = 0 \quad \text{(diffusion equation)},$$

$$u_{tt} - c^2 u_{xx} = 0 \quad \text{(wave equation)},$$

$$u_{xx} + u_{yy} = 0 \quad \text{(Laplace's equation)}.$$

The first two are *evolution equations* that describe how a process evolves in time; we would expect that each would require initial data that give the state of the system at time $t = 0$. The third is an equilibrium equation where time is not an issue—we would expect that boundary conditions are appropriate. Moreover, we suspect that solutions of these three equations would behave differently. Diffusion tends to smear out signals, whereas waves tend to propagate with some coherence; solutions to Laplace's equation are steady-state and do not vary in time at all. Therefore, if we were to classify these three equations physically, we would call them diffusive, wave-like, and static, respectively.

In general, let us consider a second-order differential equation of the form

$$Au_{xx} + Bu_{xt} + Cu_{tt} + F(x, t, u, u_x, u_t) = 0, \tag{1.52}$$

where A, B, and C are constants. Note that the second-order derivatives are assumed to appear linearly, or to the first degree; the expression $Lu \equiv Au_{xx} + Bu_{xt} + Cu_{tt}$ is called the *principal part* of the equation, and it is on this part that we base the classification. Observe that we have used the independent variables x and t; should the problem involve only spatial variables, we would replace t by y. The classification is based on the sign of the quantity

$$D \equiv B^2 - 4AC,$$

which is called the *discriminant*. If $D > 0$, we say that (1.52) is *hyperbolic*; if $D < 0$, we say that (1.52) is *elliptic*; if $D = 0$, we say that (1.52) is *parabolic*. Under this classification, the diffusion equation is parabolic, the wave equation is hyperbolic, and Laplace's equation is elliptic. This terminology comes from the familiar classification of plane curves; for example, $Ax^2 + Ct^2 = 1$, where $A, C > 0$, $B = 0$ (and so $D < 0$) graphs as an ellipse in the xt-plane. Similarly, $Ax^2 - Ct^2 = 1$ graphs as a hyperbola, and so forth. As it turns out, all parabolic equations are diffusion-like, all hyperbolic equations are wave-like, and all elliptic equations are static. Later we shall show why this is true.

If A, B, and C are not constants, but rather functions of the independent variables x and t, then the discriminant D depends on x and t. We make the same definitions in this case, but now the sign of D can change, depending upon the domain. Problems that change type from one region to another can be difficult, and we do not consider them in this text. Classification of systems of PDEs is also possible, as well as classification of PDEs with nonlinear principal parts where the coefficients A, B, and C depend on the solution u itself. The reader is referred to a more advanced treatment for a discussion of these matters.

Now we show that the principal part Lu in equation (1.52) can be simplified for each of the three types by introducing a new set of independent variables. This strategy of searching for a change of variables that simplifies a problem is common in differential equations. We examine the case where A, B, and C are constant, and we seek a linear transformation

$$\xi = ax + bt, \quad \tau = cx + dt$$

that simplifies Lu. Here ξ and τ are new independent variables, and a, b, c, and d are to be determined. We assume that $ad - bc \neq 0$, so that the transformation is invertible, that is, we may solve for x and t in terms of ξ and τ. The dependent function u in the new variables will be denoted by $U = U(\xi, \tau)$; that is, $u(x, t) = U(ax + bt, cx + dt)$. Then, by the chain rule for derivatives,

$$u_x = U_\xi \xi_x + U_\tau \tau_x = aU_\xi + cU_\tau,$$

$$u_t = U_\xi \xi_t + U_\tau \tau_t = bU_\xi + dU_\tau.$$

To get the second partial requires another application of the chain rule. We have

$$\begin{aligned} u_{xx} &= \frac{\partial}{\partial x}(aU_\xi + cU_\tau) \\ &= a(aU_{\xi\xi} + cU_{\xi\tau}) + c(aU_{\xi\tau} + cU_{\tau\tau}) \\ &= a^2 U_{\xi\xi} + 2acU_{\xi\tau} + c^2 U_{\tau\tau}. \end{aligned}$$

The other two second partial derivatives are computed similarly, and we get

$$u_{tt} = b^2 U_{\xi\xi} + 2bdU_{\xi\tau} + d^2 U_{\tau\tau},$$

$$u_{xt} = abU_{\xi\xi} + (ad + cb)U_{\xi\tau} + cdU_{\tau\tau}.$$

Substituting these quantities into the principal part and collecting terms, we get

$$\begin{aligned} Au_{xx} + Bu_{xt} + Cu_{tt} = {} & (Aa^2 + Bab + Cb^2)U_{\xi\xi} \\ & + (2acA + B(ad + bc) + 2Cbd)U_{\xi\tau} \\ & + (Ac^2 + Bcd + Cd^2)U_{\tau\tau}. \end{aligned} \tag{1.53}$$

Now we can select a, b, c, d so that some of the second partials in the new variables disappear. This process must be handled differently depending on the sign of the discriminant D.

Hyperbolic case, $D > 0$

We have some flexibility, so let us choose $a = c = 1$. Then the coefficients of $U_{\xi\xi}$ and $U_{\tau\tau}$, which have the same form, become quadratic expressions in b and d, respectively, and we can force those coefficients to vanish by choosing

$$b = \frac{-B + \sqrt{D}}{2C}, \quad d = \frac{-B - \sqrt{D}}{2C}. \tag{1.54}$$

Here we have just used the quadratic formula. The remaining coefficient, that of $U_{\xi\tau}$, is nonzero (we leave this as an exercise). In summary, the transformation

$$\xi = x + \left(\frac{-B + \sqrt{D}}{2C} \right) t, \quad \tau = x + \left(\frac{-B - \sqrt{D}}{2C} \right) t \tag{1.55}$$

transforms the PDE (1.52) into a simpler equation of the form

$$U_{\xi\tau} + G(\xi, \tau, U, U_\xi, U_\tau) = 0,$$

where only the mixed second partial derivative appears. Thus there is a significant simplification over (1.55), where all the second partial derivatives occur.

This latter equation is called the *canonical form* of a hyperbolic equation, and the coordinates ξ and τ defined by (1.55) are called the *characteristic coordinates*. Transformation to these coordinates is almost always a preferred strategy for hyperbolic PDEs.

Finally, if $C = 0$, then (1.54) is not valid; in this case select $b = d = 1$ and

$$a = \frac{-B + \sqrt{D}}{2A}, \quad c = \frac{-B - \sqrt{D}}{2A}.$$

Parabolic case, $D = 0$

Now equations (1.54) give $b = d$, and the resulting transformation $\xi = x + bt$, $\tau = x + bt$ is not invertible. So we must proceed differently. Observe that if we choose $a = c = 1$, $d = -B/(2C)$, and $b = 0$, then now the coefficients of $U_{\tau\tau}$ and $U_{\xi\tau}$ in (1.53) vanish (another exercise!). Therefore, the transformation

$$\xi = x, \quad \tau = x - \frac{B}{2C} t$$

transforms equation (1.52) into

$$U_{\xi\xi} + H(\xi, \tau, U, U_\xi, U_\tau) = 0,$$

where only a double partial appears, and this is the *canonical form* in the parabolic case. Observe its similarity to the heat equation.

Elliptic case, $D < 0$

Now b and d in (1.54) are conjugate complex numbers, i.e., $d = \bar{b}$. Selecting $a = c = 1$, we obtain a complex transformation

$$\xi = x + bt, \quad \tau = x + \bar{b}t.$$

But then, a real transformation can be found by taking real variables

$$\alpha = \frac{1}{2}(\xi + \tau), \quad \beta = \frac{1}{2i}(\xi - \tau).$$

It is an exercise to show that this transforms (1.49) into the equation

$$U_{\alpha\alpha} + U_{\beta\beta} + K(\alpha, \beta, U, U_\alpha, U_\beta) = 0,$$

where both double partials are present and the mixed partial is absent; the last equation is the *canonical form* for elliptic equations. We recognize the combination of second partial derivatives as the Laplacian operator.

Generally, characteristic coordinates are useful only for hyperbolic equations, and they do not play a particularly important role in elliptic and parabolic equations. First-order equations, like a reaction–advection equation, are classified as hyperbolic because they propagate signals like the wave equation. Second-order linear homogeneous equations with constant coefficients are also sometimes classified by their dispersion relation $\omega = \omega(k)$ (see Exercise 10, Section 1.1). If $\omega(k)$ is complex, the PDE is called *diffusive*, and if $\omega(k)$ is real and $\omega''(k) \neq 0$, the PDE is called *dispersive*. The diffusion equation is diffusive; the wave equation is neither diffusive nor dispersive. The term dispersive means that the speed $\omega(k)/k$ a plane wave $u = Ae^{i(kx-\omega(k)t)}$ travels depends upon the wave number k. So waves of different wavelength travel at different speeds, and thus "disperse."

Exercises

1. Classify the PDE

$$u_{xx} + 2ku_{xt} + k^2 u_{tt} = 0, \quad k \neq 0.$$

Find a transformation $\xi = x + bt$, $\tau = x + dt$ of the independent variables that transforms the equation into a simpler equation of the form $U_{\xi\xi} = 0$. Find the solution to the given equation in terms of two arbitrary functions.

2. Find the solution of the PDE

$$2u_{xx} - 4u_{xt} + u_x = 0$$

in terms of two arbitrary functions. Hint: Make a transformation.

3. Classify the PDE

$$xu_{xx} - 4u_{xt} = 0$$

in the region $x > 0$. Observe that the PDE has variable coefficients. Solve the equation by making the *nonlinear* change of variables $\xi = t$, $\tau = t + 4\ln x$. The solution is

$$u = e^{-t/4}f(t + 4\ln x) + g(t),$$

where f and g are arbitrary functions.

4. Show that the equation

$$u_{tt} - c^2 u_{xx} + au_t + bu_x + du = f(x, t)$$

can be transformed into an equation of the form

$$w_{\xi\tau} + kw = g(\xi, \tau), \quad w = w(\xi, \tau)$$

by first making the transformation $\xi = x - ct$, $\tau = x + ct$ and then letting $u = we^{\alpha\xi + \beta\tau}$ for some choice of α, β.

5. Classify the PDE

$$u_{xx} - 6u_{xy} + 12u_{yy} = 0.$$

Find a transformation of independent variables that changes it to Laplace's equation.

6. Classify the PDE

$$x^2 u_{xx} + 2xu_{xt} + u_{tt} = u_t$$

and find the general solution using the transformation $\xi = xe^{-t}$, $\tau = t$.

7. Show that the diffusion equation is diffusive and the wave equation is neither diffusive nor dispersive. (See Exercise 10, Section 1.1).

CHAPTER 2 Partial Differential Equations on Unbounded Domains

2.1 Cauchy Problem for the Heat Equation

In this chapter we investigate problems on unbounded spatial domains. We begin with the heat equation, or diffusion equation, on the real line. That is, we consider the initial value problem

$$u_t = ku_{xx}, \quad x \in \mathbf{R}, \; t > 0, \tag{2.1}$$

$$u(x, 0) = \phi(x), \quad x \in \mathbf{R}. \tag{2.2}$$

Physically, this problem is a model of heat flow in an infinitely long bar where the initial temperature $\phi(x)$ is prescribed. In a chemical or biological context, the equation governs density variations under a diffusion process. Notice that there are no boundaries in the problem, so we do not prescribe boundary conditions explicitly. However, for problems on infinite domains, conditions at infinity are sometimes either stated explicitly or understood. Such a condition might require boundedness of the solution or some type of decay of the solution to zero as $x \to \pm\infty$. In mathematics, a pure initial value problem like (2.1)–(2.2) is often called a *Cauchy problem*.

Deriving the solution of (2.1)–(2.2) is accomplished in two steps. First we will solve the problem for a special step function $\phi(x)$, and then we will construct the solution to (2.1)–(2.2) using that special solution. So, first let us consider the problem

$$w_t = kw_{xx}, \quad x \in \mathbf{R}, \ t > 0, \tag{2.3}$$

$$w(x, 0) = 0 \quad \text{for } x < 0; \quad w(x, 0) = u_0 \quad \text{for } x > 0, \tag{2.4}$$

where we have taken the initial condition to be a step function with jump u_0.

We motivate our approach to the solution of (2.3)–(2.4) with a simple idea from the subject of dimensional analysis. Dimensional analysis deals with the study of units (seconds, meters, kilograms, and so forth) and dimensions (time, length, mass, and so forth) of the quantities in a problem and how they relate to each other. Equations must be dimensionally consistent (one cannot add apples to oranges), and important conclusions can be drawn from this fact. The cornerstone result in dimensional analysis is called the *pi theorem*. The pi theorem guarantees that whenever there is a physical law relating dimensioned quantities $q_1, \ldots, q_m$, then there is an equivalent physical law relating the independent dimensionless quantities that can be formed from $q_1, \ldots, q_m$. By a dimensionless quantity we mean one in which all the dimensions (length, time, mass, etc.) cancel out. As a simple example take the law

$$h = -\frac{1}{2}gt^2 + vt$$

that gives the height h of an object at time t when it is thrown upward with initial velocity v; the constant g is the acceleration due to gravity. Here the dimensioned quantities are h, t, v, and g, having dimensions length, time, length per time, and length per time-squared. This law can be rearranged and written equivalently as

$$\frac{h}{vt} = -\frac{1}{2}\left(\frac{gt}{v}\right) + 1$$

in terms of the two dimensionless quantities

$$\pi_1 \equiv \frac{h}{vt} \quad \text{and} \quad \pi_2 \equiv \frac{gt}{v}.$$

For example, h is a length and vt, a velocity times a time, is also a length; so π_1, or h divided by vt, has no dimensions. Similarly, $\pi_2 = gt/v$ is dimensionless. A law in dimensioned variables can always be reformulated in dimensionless quantities. So the physical law can be written as $\pi_1 = -\frac{1}{2}\pi_2 + 1$.

We use similar reasoning to guess the form of the solution of the initial value problem (2.3)–(2.4). First we list all the variables and constants in the problem: x, t, w, u_0, k. These have dimensions length, time, degrees, degrees, and length-squared per time, respectively. We notice that w/u_0 is a dimensionless quantity (degrees divided by degrees); the only other independent dimensionless quantity in the problem is $x/\sqrt{4kt}$ (the "4" is included for convenience). We expect, therefore, that the solution can be written as some combination of these dimensionless variables, or

$$\frac{w}{u_0} = f(\frac{x}{\sqrt{4kt}})$$

for some function f to be determined. In fact, this is the case. So let us substitute

$$w = f(z), \quad z = \frac{x}{\sqrt{4kt}}$$

into the PDE (2.3). We have taken $u_0 = 1$ for simplicity. The chain rule allows us to compute the partial derivatives as

$$w_t = f'(z)z_t = -\frac{1}{2}\frac{x}{\sqrt{4kt^3}}f'(z),$$

$$w_x = f'(z)z_x = \frac{1}{\sqrt{4kt}}f'(z),$$

$$w_{xx} = \frac{\partial}{\partial x}w_x = \frac{1}{4kt}f''(z).$$

Substituting these quantities into (2.3) then gives, after some cancellation, an ordinary differential equation,

$$f''(z) + 2zf'(z) = 0,$$

for $f(z)$. This equation is easily solved by multiplying through by the integrating factor e^{z^2} and integrating to get

$$f'(z) = c_1 e^{-z^2},$$

where c_1 is a constant of integration. Integrating from 0 to z gives

$$f(z) = c_1 \int_0^z e^{-r^2} dr + c_2,$$

where c_2 is another constant of integration. Therefore we have determined solutions of (2.3) of the form

$$w(x,t) = c_1 \int_0^{x/\sqrt{4kt}} e^{-r^2} dr + c_2.$$

Next we apply the initial condition (2.4) (with $u_0 = 1$) to determine the constants c_1 and c_2. For a fixed $x < 0$ we take the limit as $t \to 0$ to get

$$0 = w(x,0) = c_1 \int_0^{-\infty} e^{-r^2} dr + c_2.$$

For a fixed $x > 0$ we take the limit as $t \to 0$ to get

$$1 = w(x,0) = c_1 \int_0^{\infty} e^{-r^2} dr + c_2.$$

Recalling that

$$\int_0^\infty e^{-r^2}\,dr = \frac{\sqrt{\pi}}{2},$$

we can solve the last two equations to get $c_1 = 1/\sqrt{\pi}$, $c_2 = 1/2$. Therefore, the solution to (2.3)–(2.4) with $u_0 = 1$ is

$$w(x,t) = \frac{1}{2} + \frac{1}{\sqrt{\pi}} \int_0^{x/\sqrt{4kt}} e^{-r^2}\,dr. \tag{2.5}$$

This solution can be written nicely as

$$w(x,t) = \frac{1}{2}\left(1 + \operatorname{erf}\left(\frac{x}{\sqrt{4kt}}\right)\right) \tag{2.6}$$

in terms of a special function called the "erf" function, which is defined by

$$\operatorname{erf}(z) = \frac{2}{\sqrt{\pi}} \int_0^z e^{-r^2}\,dr.$$

Figure 2.1 shows a graph of several time snapshots of the solution (2.6).

Now we will use (2.5) and a physical argument to deduce a solution to the Cauchy problem (2.1)–(2.2). Later, in Section 2.7, we present an analytical argument based on Fourier transforms. We make some observations. First, if a function w satisfies the heat equation, then so does w_x, the partial derivative of that function with respect to x. This is easy to see because

$$0 = (w_t - kw_{xx})_x = (w_x)_t - k(w_x)_{xx}.$$

Therefore, since $w(x,t)$ solves the heat equation, the function

$$G(x,t) \equiv w_x(x,t)$$

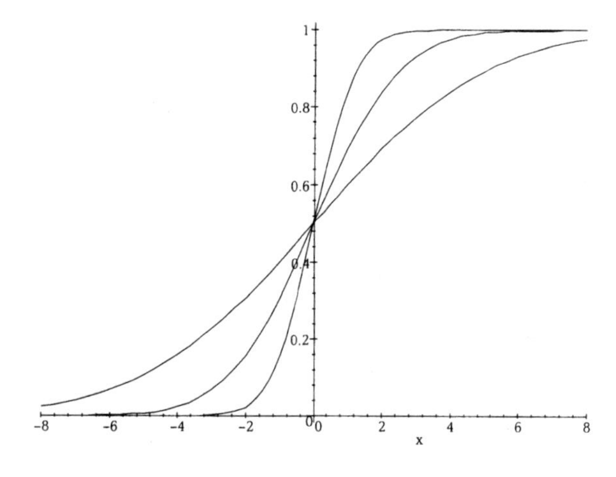

Figure 2.1. Temperature profiles for different times when the initial temperature is a step function.

solves the heat equation. By direct differentiation we find that

$$G(x,t) = \frac{1}{\sqrt{4\pi kt}} e^{-x^2/(4kt)}. \tag{2.7}$$

The function G is called the *heat kernel* or *fundamental solution* to the heat equation; the reader will note that for each $t > 0$ it graphs as a bell-shaped curve (see Exercise 1, Section 1.1), and the area under the curve for each $t > 0$ is one; that is,

$$\int_{-\infty}^{\infty} G(x,t)dx = 1, \quad t > 0.$$

$G(x,t)$ is the temperature surface that results from an initial unit heat source, i.e., injecting a unit amount of heat at $x = 0$ at time $t = 0$. We further observe that shifting the temperature profile again leads to a solution to the heat equation. Thus, $G(x-y,t)$, which is the temperature surface caused by an initial unit heat source at y, solves the heat equation for any fixed, but arbitrary, y. If $\phi(y)$, rather than unity, is the magnitude of the source at y, then $\phi(y)G(x-y,t)$ gives the resulting temperature surface; the area under a temperature profile is now $\phi(y)$, where y is the location of the source. Now, let us regard the initial temperature function ϕ in (2.2) as a continuous distribution of sources $\phi(y)$ for each $y \in \mathbf{R}$. Then, superimposing all the effects $\phi(y)G(x-y,t)$ for all y gives the total effect of all these isolated sources; that is,

$$\begin{aligned} u(x,t) &= \int_{-\infty}^{\infty} \phi(y)G(x-y,t)dy \\ &= \int_{-\infty}^{\infty} \phi(y)\frac{1}{\sqrt{4\pi kt}} e^{-(x-y)^2/(4kt)}dy \end{aligned} \tag{2.8}$$

is a solution to the Cauchy problem (2.1)–(2.2) for reasonable assumptions on the initial condition ϕ. More precisely, if ϕ is a bounded continuous function, then $u(x,t)$ given by (2.8) is a solution to the heat equation and $u(x,t) \to \phi(x)$ as $t \to 0^+$. If ϕ is only piecewise continuous (i.e., it has finitely many jump discontinuities in any bounded interval $[a,b]$ in $\mathbf{R}$), then $u(x,t)$ still solves (2.1), but as $t \to 0^+$ the solution approaches the average value of the left and right limits at a point of discontinuity of ϕ; in symbols, $u(x,t) \to \frac{1}{2}(\phi(x^-) + \phi(x^+))$ as $t \to 0^+$. (A precise definition of piecewise continuity is given in Section 3.3).

This discussion of the Cauchy problem for the heat equation has been intuitive, and it is a basis for understanding why the solution has the form it does.

There is another way to write the solution (2.8) to the Cauchy problem (2.1)–(2.2). If we change variables in the integral using the substitution

$r = (x - y)/\sqrt{4kt}$, then $dr = -dy/\sqrt{4kt}$, and (2.8) becomes

$$u(x, t) = \frac{1}{\sqrt{\pi}} \int_{-\infty}^{\infty} e^{-r^2} \phi(x - r\sqrt{4kt})dr. \tag{2.9}$$

This formula is called the *Poisson integral representation*.

We make several observations. First, the solution (2.8) or (2.9) of the Cauchy problem is an integral representation; although the formula is not complicated, for most initial conditions $\phi(x)$ the integration cannot be performed analytically. Therefore, numerical or computer evaluation of the integral will ultimately be required if temperature profiles are desired. Also, notice that the temperature $u(x, t)$ is nonzero for every real x, even if ϕ is zero outside a small interval about the origin. Thus, a signal propagated by the heat, or diffusion, equation travels infinitely fast; according to this model, if odors diffuse, a bear would instantly smell a newly opened can of tuna ten miles away. Next, although we will not prove it, the solution given by (2.8) is very smooth; that is, u in infinitely differentiable in both x and t in the domain $t > 0$; this is true even if ϕ is piecewise continuous. Initial signals propagated by the heat equation are immediately smoothed out.

Finally, we note that the heat kernel $G(x, t)$ defined in (2.7) is also called the *Green's function* for the Cauchy problem. In general, the Green's function for a problem is the response of a system, or the effect, caused by a point source. In heat flow on the real line, $G(x, t)$ is the response, i.e., the temperature surface caused by a unit, point heat source given to the system at $x = 0,\ t = 0$. Some of the references discuss the construction of Green's functions for a variety of problems. Because of the basic role this function plays in diffusion problems, $G(x, t)$ is also called the *fundamental solution* to the heat equation. The reader can consult Section 1.4 for a discussion of the origin of the fundamental solution from a probability discussion.

Exercises

1. Solve the Cauchy problem (2.1)–(2.2) for the following initial conditions.

   (a) $\phi(x) = 1$ if $|x| < 1$ and $\phi(x) = 0$ if $|x| > 1$.

   (b) $\phi(x) = e^{-x}$, $x > 0$; $\phi(x) = 0$, $x < 0$.

   Write the solutions in terms of the erf function. In (b) complete the square with respect to y in the exponent of e to obtain a nice form.

2. If $|\phi(x)| \leq M$ for all x, where M is a positive constant, show that the solution u to the Cauchy problem (2.1)–(2.2) satisfies $|u(x, t)| \leq M$ for all x and $t > 0$.

Hint: Use the calculus fact that the absolute value of an integral is less than or equal to the integral of the absolute value: $|\int f| \le \int |f|$.

3. Consider the problem (2.3)–(2.4) with $u_0 = 1$. For a fixed $x = x_0$, what is the approximate temperature $w(x_0, t)$ for very large t? Hint: Expand the integrand in the formula for the solution in a power series and integrate term by term.

4. Show that if $u(x,t)$ and $v(x,t)$ are any two solutions to the heat equation (2.1), then $w(x,y,t) = u(x,t)v(y,t)$ solves the two-dimensional heat equation $w_t = k(w_{xx}+w_{yy})$. Can you guess the solution to the two-dimensional Cauchy problem

$$w_t = k(w_{xx} + w_{yy}), \quad (x,y) \in \mathbf{R}^2, \ t > 0,$$

$$w(x,y,0) = \psi(x,y), \quad (x,y) \in \mathbf{R}^2?$$

5. Let the initial temperature in the Cauchy problem (2.1)–(2.2) be given by $\phi(x) = e^{-|x+2|} + e^{-|x-2|}$, with $k = 1$. Use the numerical integration operation in a computer algebra package to draw temperature profiles at several times to illustrate how heat flows in this system. Exhibit the temperature profiles on a single set of coordinate axes.

6. Verify that

$$\int_{-\infty}^{\infty} G(x,t)dx = 1, \quad t > 0.$$

Hint: Change variables as in the derivation of Poisson's integral representation.

2.2 Cauchy Problem for the Wave Equation

The one-dimensional wave equation is

$$u_{tt} - c^2 u_{xx} = 0. \tag{2.10}$$

We observed in Section 1.5 that it models the amplitude of waves on a string and small amplitude signals in acoustics. It also arises in electromagnetic wave propagation, in the mechanical vibrations of elastic media, as well as in other problems. It is a hyperbolic equation and is one of the three fundamental equations in PDEs (along with the diffusion equation and Laplace's equation). Under the transformation of variables (to characteristic coordinates)

$$\xi = x - ct, \qquad \tau = x + ct,$$

the wave equation is transformed into the canonical form

$$U_{\tau\xi} = 0, \quad U = U(\xi, \tau),$$

which can be integrated twice to obtain the general solution

$$U(\xi, \tau) = F(\xi) + G(\tau),$$

where F and G are arbitrary functions. Thus, the general solution to (2.10) is

$$u(x, t) = F(x - ct) + G(x + ct). \tag{2.11}$$

Hence, solutions of the wave equation are the superposition (sum) of right- and left-traveling waves moving at speed c.

The Cauchy problem for the wave equation is

$$u_{tt} - c^2 u_{xx} = 0, \quad x \in \mathbf{R}, \ t > 0, \tag{2.12}$$

$$u(x, 0) = f(x), \quad u_t(x, 0) = g(x), \quad x \in \mathbf{R}. \tag{2.13}$$

Here, f defines the initial displacement of an infinite string, and g defines its initial velocity. The equation is second-order in t, so both the position and velocity must be specified initially.

There is a simple analytical formula for the solution to the Cauchy problem (2.12)–(2.13). It is called *d'Alembert's formula*, and it is given by

$$u(x, t) = \frac{1}{2}(f(x - ct) + f(x + ct)) + \frac{1}{2c}\int_{x-ct}^{x+ct} g(s)ds. \tag{2.14}$$

If f'' and g' are continuous, then it is a straightforward exercise in differential calculus, using Leibniz's formula, to verify that this formula solves (2.12)–(2.13). The formula can be derived (see Exercise 1) by determining the two functions F and G in (2.11) using the initial data (2.13).

Much insight into the behavior of solutions comes from examining the special case where the initial velocity is zero and the initial displacement is a bell-shaped curve. Specifically, we consider the problem (with $c = 2$)

$$u_{tt} - 4u_{xx} = 0, \quad x \in \mathbf{R}, \ t > 0,$$

$$u(x, 0) = e^{-x^2}, \quad u_t(x, 0) = 0, \quad x \in \mathbf{R}.$$

The exact solution is, by d'Alembert's formula,

$$u(x, t) = \frac{1}{2}(e^{-(x-2t)^2} + e^{-(x+2t)^2}).$$

Either the solution surface or wave profiles can be graphed easily using a computer algebra package. Figure 2.2 shows the solution surface; observe how the initial signal splits into two smaller signals, and those travel off in opposite directions at speed $c = 2$. In the exercises the reader is asked to examine the case where $f = 0$ and $g \neq 0$; this is the case where the initial displacement is zero and the string is given an initial velocity, or impulse, by, say, striking the string with an object.

Close examination of d'Alembert's formula reveals a fundamental property of the wave equation. If the initial disturbance is *supported* in some

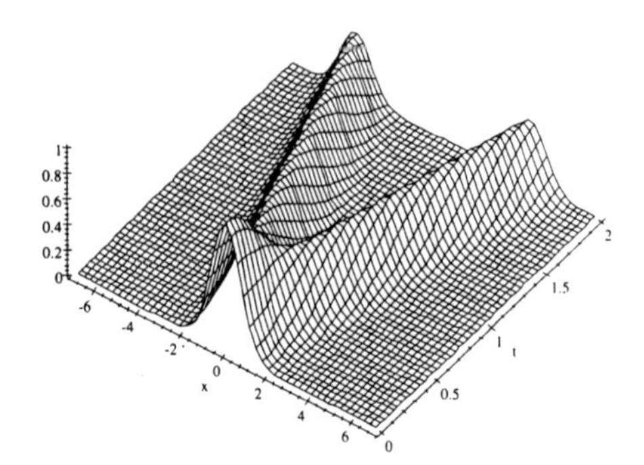

Figure 2.2. The solution surface.

interval $a \leq x \leq b$ (this means that it is zero outside that interval, so the signal is located only in $a \leq x \leq b$), then the signal is always zero outside the region bounded by the two straight lines $x + ct = a$ and $x - ct = b$. See Figure 2.3. This region is called the *region of influence* of the interval $[a, b]$. An initial signal in $[a, b]$ can never affect the solution outside this region. The lines $x + ct =$ constant are paths in space–time along which signals are propagated at velocity $-c$; the lines $x - ct = constant$ are paths in space–time along which signals are propagated with velocity c. These two families of lines are called the *negative* and *positive characteristics*, respectively. If the interval $[a, b]$ is shrunk to a point, then the region of influence shrinks to a cone, which is called the *light cone*. Looking at the situation in reverse, we can ask what initial data can affect the solution at a point (x_0, t_0). From the d'Alembert formula, only the initial values in the interval $[x_0 - ct_0, x_0 + ct_0]$ will affect the solution at (x_0, t_0). This interval is called the *domain of dependence*, and it is found by tracing the characteristics emanating from the point (x_0, t_0) backward in time to the x-axis.

In summary, there are important points to note regarding the characteristic curves. First, they are curves that carry the signals forward in space–time with velocity c and $-c$. Second, they define a special coordinate system $\xi = x - ct$, $\tau = x + ct$ under which the wave equation $u_{tt} - c^2 u_{xx} = 0$ is reduced to the simple canonical form $u_{\xi\tau} = 0$. In hy-

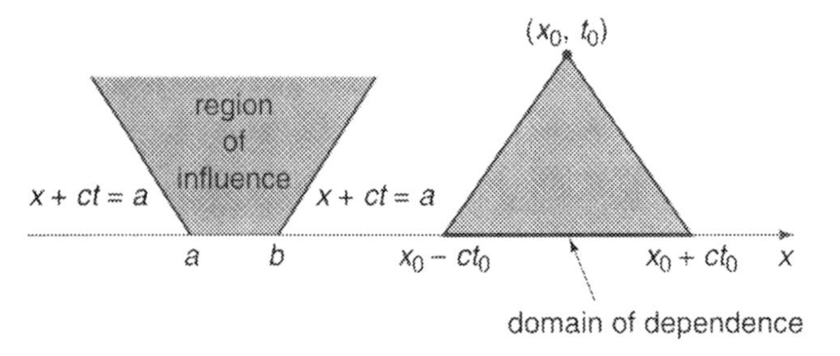

Figure 2.3. Region of influence and domain of dependence.

perbolic problems there is always a set of characteristic curves that play these roles. Even first order PDEs, which are actually wave-like, have one family of such curves that carry signals and provide a distinguished coordinate sytem where the problem simplifies (recall the examples in Section 1.2).

Finally, we point out again the important differences between parabolic and hyperbolic problems. Hyperbolic, or wave-like, equations propagate signals at a finite speed along characteristics; there is coherency in the wave form as it propagates, and therefore information is retained. Parabolic, or diffusion, equations propagate signals at infinite speed; because the signals diffuse or smear out, there is a gradual loss of information. A good way to understand how different equations propagate information is to determine how a signal is propagated in a special case. For example, suppose the initial signal is a Gaussian function or bell-shaped curve $\exp(-x^2)$. Think of this signal as being a bit of information. The convection equation $u_t + cu_x = 0$, which is a wave-like equation, propagates this signal via

$$u(x, t) = e^{-(x-ct)^2}.$$

That is, it moves it at speed c without distortion. The wave equation $u_{tt} = c^2 u_{xx}$ moves it via

$$u(x, t) = 0.5(e^{-(x-ct)^2} + e^{-(x+ct)^2}).$$

So the signal breaks into two pieces, and they propagate in opposite directions at speed c. The diffusion equation $u_t = ku_{xx}$ propagates the signal via

$$u(x, t) = \frac{1}{\sqrt{1 + 4kt}} e^{-x^2/(1+4kt)}.$$

So the signal stays at the same place, but it spreads out and decreases in amplitude. Any information in the signal is eventually lost.

Exercises

1. Derive d'Alembert's formula (2.14) by determining the two arbitrary functions F and G in the general solution (2.11) using the initial conditions (2.13).

2. Calculate the exact solution to the Cauchy problem when $c = 2$, the initial displacement is $f(x) = 0$, and the initial velocity is $g(x) = 1/(1 + 0.25x^2)$. Plot the solution surface and discuss the effect of giving a string at rest an initial impulse. Contrast the solution with the case when $f \neq 0$ and $g = 0$.

3. Solve the outgoing signal problem

$$u_{tt} - c^2 u_{xx} = 0, \quad x > 0, \ -\infty < t < \infty; \quad u_x(0, t) = s(t), \quad -\infty < t < \infty,$$

where $s(t)$ is a known signal. Hint: Look for a right-traveling wave solution.

4. The three-dimensional wave equation is

$$u_{tt} - c^2 \Delta u = 0,$$

where $u = u(x, y, z, t)$ and Δ is the Laplacian operator. For waves with spherical symmetry, $u = u(\rho, t)$, where $\rho = \sqrt{x^2 + y^2 + z^2}$. In this special case the Laplacian is given by (Section 1.8) $\Delta u = u_{\rho\rho} + \frac{2}{\rho} u_\rho$. By introducing a change of dependent variable $U = \rho u$, show that the general solution for the spherically symmetric wave equation

$$u_{tt} = c^2 (u_{\rho\rho} + \frac{2}{\rho} u_\rho)$$

is

$$u = \frac{1}{\rho}(F(\rho - ct) + G(\rho + ct)).$$

Why do you think an outward-moving wave $u = F(\rho - ct)/\rho$ decays in amplitude?

5. Solve the Cauchy problem

$$u_{tt} - c^2 u_{xx} = 0, \quad x \in \mathbf{R}, \ t > 0,$$

$$u(x, 0) = e^{-|x|}, \quad u_t(x, 0) = \cos x, \quad x \in \mathbf{R}.$$

Use a computer algebra program to graph the wave profile at $t = 1, 2, 3$. Take $c = 1$.

6. In Section 1.7 we showed that any solution to Laplace's equation has the property that its value at a point is approximately the average of four nearby values surrounding the point. Can we make a statement about solutions to the wave equation? Consider any characteristic parallelogram (see Figure 2.4) whose sides are positive and negative characteristics, and let A, B, C, D be the vertices as shown. Show that any solution to the wave equation satisfies the relation

$$u(A) + u(C) = u(B) + u(D).$$

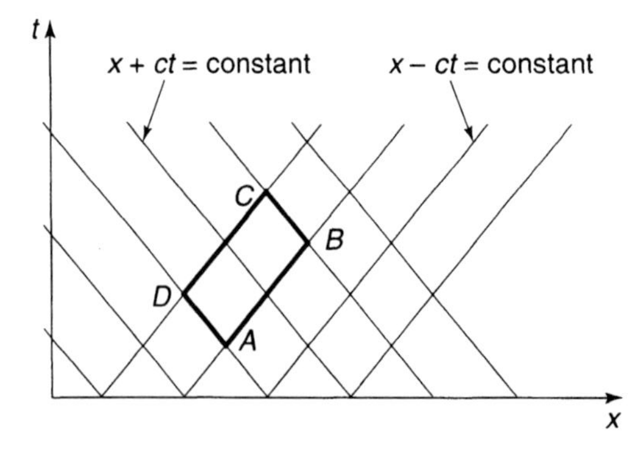

Figure 2.4. Characteristic parallelogram.

7. Let $u = u(x, t)$ solve the wave equation; show that $v = v(x, t)$ defined by

$$v(x, t) = \frac{c}{\sqrt{4\pi kt}} \int_{-\infty}^{\infty} u(s, x) e^{-s^2 c^2/(4kt)} ds.$$

solves the heat equation.

2.3 Ill-Posed Problems

In the last two sections we solved the Cauchy problem, or pure initial value problem, for the heat equation and for the wave equation. Now let us set up a similar problem for Laplace's equation. Right away the reader should be skeptical because Laplace's equation is an elliptic equilibrium equation that does not involve time; we expect boundary conditions rather than initial conditions. Nevertheless, let us consider the two-dimensional Laplace's equation in the upper-half plane $y > 0$ and prescribe u *and* its y derivative u_y along $y = 0$, or the x axis. That is, let us consider the problem

$$u_{xx} + u_{yy} = 0, \quad x \in \mathbf{R}, y > 0, \tag{2.15}$$

$$u(x, 0) = f(x), \quad u_y(x, 0) = g(x), \quad x \in \mathbf{R}. \tag{2.16}$$

Notice the similarity to the initial value problem for the wave equation (2.12)–(2.13); but here we are using y instead of t, and the equation is elliptic rather than hyperbolic. Note also that (2.15)–(2.16) is neither a Dirichlet problem nor a Neumann problem because both u and u_y are specified along the boundary $y = 0$.

Now let us analyze this problem carefully in a special case. Suppose first that

$$f(x) = g(x) = 0.$$

Then it is clear that a solution (the only solution) to (2.15)–(2.16) is the zero solution $u(x, y) \equiv 0$ for all $x \in \mathbf{R}$, $y > 0$. Next let us change the conditions along the boundary by taking

$$f(x) = \frac{1}{n} \cos nx, \quad g(x) = 0, \quad x \in \mathbf{R}.$$

Then, as one can easily check, the solution to (2.15)–(2.16) is

$$u(x, y) = \frac{1}{n} \cos nx \cosh ny.$$

Suppose n is large; then we have changed the boundary condition $u(x, 0) = f(x)$ by only a small amount. Yet the solution has changed from zero by a large amount; for example, along the line $x = 0$ the solution is

$$u(0, y) = \frac{\cosh ny}{n} \to +\infty \quad \text{as } y \to \infty.$$

Consequently, in this example, a small change of the data on the boundary produced a large change in the solution. This behavior is disturbing, because in a physical problem we expect that the solution should depend continuously on the boundary data—a small change in the boundary data should produce a small change in the solution. After all, we want to be confident in the accuracy of our solution even if we get the boundary data only approximately correct. This latter property of continuous dependence on data is called *stability*, and the problem (2.15)–(2.16) does not have it. So the Cauchy problem for Laplace's equation (2.15)–(2.16) is not a well-posed problem. In fact, since Laplace's equation models steady heat flow, it seems physically reasonable that we need only specify the temperature $u(x, 0)$ along the boundary to be able to solve the problem, or specify the flux $u_y(x, 0)$, but not both as in (2.16).

The term *well-posed* in PDEs has a technical meaning. We say that a boundary value problem, initial value problem, or an initial boundary value problem is well-posed if (i) it has a solution, (ii) the solution is unique, and (iii) the solution depends continuously on the initial and/or boundary data (stability). Resolving these three questions for various PDE models occupies much of the theory of PDEs.

Now that we have shown that the Cauchy problem for Laplace's equation is not stable, the reader may be skeptical about other problems such as the Cauchy problem for the heat equation. We can easily observe that solutions to this problem have the stability property. To this end, consider the two problems

$$u_t = ku_{xx}, \quad x \in \mathbf{R}, t > 0,$$

$$u(x, 0) = \phi(x), \quad x \in \mathbf{R},$$

and

$$v_t = kv_{xx}, \quad x \in \mathbf{R}, t > 0,$$

$$v(x, 0) = \psi(x), \quad x \in \mathbf{R},$$

where ϕ and ψ are continuous, bounded functions and close in the sense that $|\phi(x) - \psi(x)| \le \delta$ for all x, where δ is a small number. We would like to show that the corresponding solutions $u(x, t)$ and $v(x, t)$ are close. Let us define $w(x, t) = u(x, t) - v(x, t)$ and note that w satisfies the Cauchy problem

$$w_t = kw_{xx}, \quad x \in \mathbf{R}, t > 0,$$

$$w(x, 0) = \phi(x) - \psi(x), \quad x \in \mathbf{R}.$$

The solution formula (2.8) gives

$$w(x,t) = \int_{-\infty}^{\infty} (\phi(y) - \psi(y))G(x-y,t)dy.$$

Therefore, for each $t > 0$,

$$|u(x,t) - v(x,t)| \le \int_{-\infty}^{\infty} | \phi(y) - \psi(y) || G(x-y,t) | dy$$
$$\le \int_{-\infty}^{\infty} \delta G(x-y,t)dy = \delta,$$

since $\int G(x-y,t)dy = 1$. Therefore, in the sense interpreted above, closeness of the initial data implies closeness of the solution.

Exercises

1. Show that the Cauchy problem for the backward diffusion equation,

$$u_t + u_{xx} = 0, \quad x \in \mathbf{R}, t > 0,$$
$$u(x,0) = f(x), \quad x \in \mathbf{R},$$

is unstable by considering the solutions

$$u(x,t) = 1 + \frac{1}{n} e^{n^2 t} \sin nx$$

for large n.

2. Let $u = u(x,y)$. Is the problem

$$u_{xy} = 0, \quad 0 < x, y < 1,$$

on the unit square, where the value of u is prescribed on the boundary of the square, a well-posed problem? Discuss.

3. Consider two Cauchy problems for the wave equation with different initial data:

$$u^i_{tt} = c^2 u^i_{xx}, \quad x \in \mathbf{R}, \quad 0 < t < T,$$
$$u^i(x,0) = f^i(x), \quad u^i_t(x,0) = g^i(x), \quad x \in \mathbf{R},$$

for $i = 1, 2$, where f^1, f^2, g^1, and g^2 are given functions (the superscripts are indices and not exponents). If for all $x \in \mathbf{R}$ we have

$$| f^1(x) - f^2(x) | \le \delta_1, \quad | g^1(x) - g^2(x) | \le \delta_2,$$

show that $| u^1(x,t) - u^2(x,t) | \le \delta_1 + \delta_2 T$ for all $x \in \mathbf{R}$, $0 < t < T$. What does this mean with regard to stability?

2.4 Semi-Infinite Domains

In Sections 2.1 and 2.2 we solved the heat equation and the wave equation, respectively, on the domain $-\infty < x < \infty$. Now we study these problems when the domain is semi-infinite, i.e., on the interval $0 < x < \infty$. This means that there is a boundary in the problem, at $x = 0$, and one expects that it is necessary to impose a boundary condition there. For example, to determine how the temperature distribution evolves in a semi-infinite bar, one should know the temperature in the bar initially, as well as the temperature at $x = 0$.

Therefore, we consider the initial boundary value problem for the heat equation

$$u_t = ku_{xx}, \quad x > 0, t > 0, \tag{2.17}$$

$$u(0,t) = 0, \quad t > 0, \tag{2.18}$$

$$u(x,0) = \phi(x), \quad x > 0, \tag{2.19}$$

where we have specified the temperature to be zero at $x = 0$ for all time. To solve this problem we use the method of *reflection* through the boundary. The idea is to extend the problem (2.17)–(2.19) to the entire real axis by extending the initial data ϕ to an *odd function* ψ defined by

$$\psi(x) = \phi(x) \;\text{ if }\; x > 0; \qquad \psi(x) = -\phi(-x) \;\text{ if }\; x < 0; \quad \psi(0) = 0.$$

We then solve the extended problem by formula (2.8) and then restrict that solution to the positive real axis, which will then be the solution to (2.17)–(2.19). Figure 2.5 shows the initial data for the extended problem and a resulting odd solution profile $v(x,t)$. Physically, we are attaching a bar occupying the space $-\infty < x < 0$ and giving it an initial temperature equal to the negative of that in the original bar.

Therefore, let us consider the Cauchy problem

$$v_t = kv_{xx}, \quad x \in \mathbf{R}, t > 0, \tag{2.20}$$

$$v(x,0) = \psi(x), \quad x \in \mathbf{R}, \tag{2.21}$$

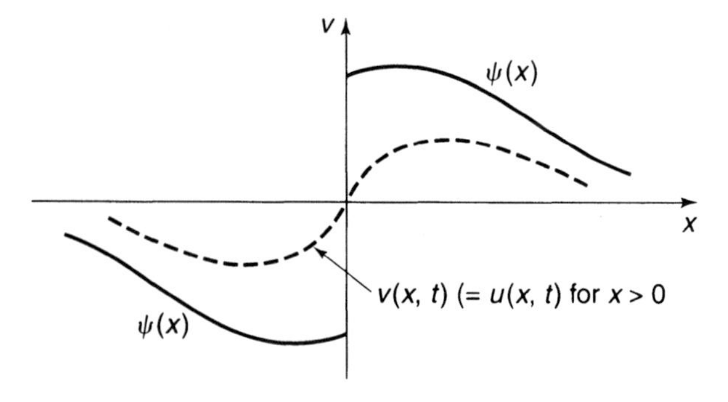

Figure 2.5. The initial data $\psi(x)$ and solution profile for the odd, extended problem.

where ψ is the odd extension of the function ϕ as defined above. By formula (2.8) the solution to (2.20)–(2.21) is given by

$$v(x,t) = \int_{-\infty}^{\infty} G(x-y,t)\psi(y)dy.$$

Breaking up this integral into two parts, $y < 0$ and $y > 0$, we obtain

$$\begin{aligned}
v(x,t) &= \int_{-\infty}^{0} G(x-y,t)\psi(y)dy + \int_{0}^{\infty} G(x-y,t)\psi(y)dy \\
&= -\int_{-\infty}^{0} G(x-y,t)\phi(-y)dy + \int_{0}^{\infty} G(x-y,t)\phi(y)dy \\
&= -\int_{0}^{\infty} G(x+y,t)\phi(y)dy + \int_{0}^{\infty} G(x-y,t)\phi(y)dy \\
&= \int_{0}^{\infty} [G(x-y,t) - G(x+y,t)]\phi(y)dy.
\end{aligned}$$

We restrict this solution to $x > 0$, and therefore the solution to (2.17)–(2.19) is

$$u(x,t) = \int_{0}^{\infty} [G(x-y,t) - G(x+y,t)]\phi(y)dy.$$

The wave equation on a semi-infinite domain can be solved in the same manner. Consider the problem of the transverse vibrations of a string occupying $x > 0$ when the end at $x = 0$ is held fixed. The initial boundary value problem is

$$u_{tt} = c^2 u_{xx}, \quad x > 0, t > 0, \tag{2.22}$$

$$u(0,t) = 0, \quad t > 0, \tag{2.23}$$

$$u(x,0) = f(x), \quad u_t(x,0) = g(x), \quad x > 0. \tag{2.24}$$

For $x > ct$ (i.e., ahead of the leading signal from the boundary) the interval of dependence lies in $(0, \infty)$, where the initial data are given; therefore, in this domain, the solution is given by d'Alembert's formula:

$$u(x,t) = \frac{1}{2}(f(x-ct) + f(x+ct)) + \frac{1}{2c}\int_{x-ct}^{x+ct} g(s)ds, \quad x > ct. \tag{2.25}$$

The data given along the $x = 0$ boundary cannot affect the solution in the region $x > ct$, since signals travel outward from the boundary at speed c. See Figure 2.6.

To solve the problem in the region $0 < x < ct$ we proceed as we did with the heat equation and extend the initial data f and g to odd functions on the entire real axis. Therefore, let us consider the problem

$$v_{tt} = c^2 v_{xx}, \quad x \in \mathbf{R}, t > 0, \tag{2.26}$$

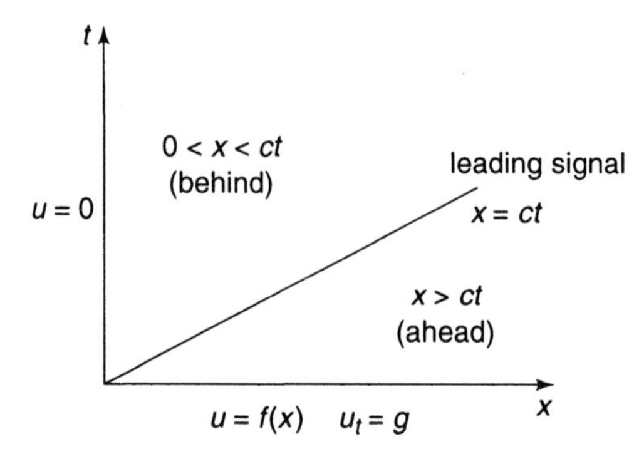

Figure 2.6. Space-time domain $x, t > 0$ where the problem (2.22)–(2.24) is defined. The region $x > ct$ is affected only by the initial data f and g and can be found by d'Alembert's formula.

$$v(0,t) = 0, \quad t > 0, \tag{2.27}$$

$$v(x,0) = F(x), \quad v_t(x,0) = G(x), \quad x \in \mathbf{R}, \tag{2.28}$$

where

$$F(x) = f(x), \quad x > 0; \quad F(0) = 0; \quad F(x) = -f(-x), \quad x < 0,$$

and

$$G(x) = g(x), \quad x > 0; \quad G(0) = 0; \quad G(x) = -g(-x), \quad x < 0.$$

By d'Alembert's formula the solution to (2.26)–(2.28) is

$$v(x,t) = \frac{1}{2}(F(x-ct) + F(x+ct)) + \frac{1}{2c}\int_{x-ct}^{x+ct} G(s)ds.$$

In the region $x < ct$ this becomes (since $x - ct < 0$)

$$v(x,t) = \frac{1}{2}(-f(-x+ct) + f(x+ct)) + \frac{1}{2c}\int_{x-ct}^{0} -g(-s)ds + \frac{1}{2c}\int_{0}^{x+ct} g(s)ds.$$

If the variable of integration s in the first integral is replaced by $-s$, then the two integrals can be combined, and we may write

$$u(x,t) = v(x,t) = \frac{1}{2}(f(x+ct) - f(ct-x)) + \frac{1}{2c}\int_{ct-x}^{ct+x} g(s)ds, \quad 0 < x < ct. \tag{2.29}$$

So the solution to the initial boundary value problem (2.22)–(2.24) is given by the two formulae (2.25) and (2.29), depending on $x > ct$ or $x < ct$.

Why does this reflection method work for the heat equation and wave equation? To reiterate, the solutions to the Cauchy problems for these two equations are odd functions if the initial data is odd. And the restriction of an odd solution to the positive real axis is the solution to the given initial boundary value problem. If this intuitive reasoning leaves the reader perplexed, then one can always verify analytically that the solutions we have obtained by this reflection method are, in fact, solutions to the given problems.

If the boundary condition (2.18) along $x = 0$ in the heat flow problem (2.17)–(2.19) is replaced by a Neumann condition

$$u_x(0,t) = 0, \quad t > 0,$$

then the problem can be solved by extending the initial data to an *even* function. The same is true for the wave equation. We leave these calculations as exercises.

Exercises

1. Solve the problem

$$u_t = ku_{xx}, \quad x > 0, t > 0,$$
$$u_x(0,t) = 0, \quad t > 0,$$
$$u(x,0) = \phi(x), \quad x > 0,$$

with an insulated boundary condition by extending ϕ to all of the real axis as an even function. The solution is

$$u(x,y) = \int_0^\infty [G(x-y,t) + G(x+y,t)]\phi(y)dy.$$

2. Find a formula for the solution to the problem

$$u_t = ku_{xx}, \quad x > 0, t > 0,$$
$$u(0,t) = 0, \quad t > 0,$$
$$u(x,0) = 1, \quad x > 0.$$

Sketch a graph of several solution profiles with $k = 0.5$.

3. Find the solution to the problem

$$u_{tt} = c^2 u_{xx}, \quad x > 0, t > 0,$$
$$u(0,t) = 0, \quad t > 0,$$
$$u(x,0) = xe^{-x}, \quad u_t(x,0) = 0, \quad x > 0.$$

Pick $c = 0.5$ and sketch several time snapshots of the solution surface to observe the reflection of the wave from the boundary.

4. Solve the problem

$$u_t = ku_{xx}, \quad x > 0, t > 0,$$
$$u(0, t) = 1, \quad t > 0,$$
$$u(x, 0) = 0, \quad x > 0.$$

Hint: Transform the problem to one of the form (2.17)–(2.19) and use Exercise 2.

5. In this exercise we use Lord Kelvin's argument, given in the mid 1860s, to estimate the age of the earth using a measurement of the geothermal gradient at the surface. The geothermal gradient is the temperature gradient u_x measured at surface of the earth. To obtain the estimate, treat the earth as flat with $x > 0$ measuring the depth from the surface $x = 0$. Assume that the diffusivity is $k = 0.007$ cm^2 per second, the initial temperature was 7000 degrees Fahrenheit (molten rock), and assume the temperature of the surrounding atmosphere has always been 0 degrees. Use a current geothermal gradient value of 3.7×10^{-4} degrees per cm. After determining the approximate age, estimate the percentage of the original heat that has been lost until the present day? Comment on the accuracy of the Kelvin's argument.

6. A clutch of insect eggs lies at a depth of x_1 cm below the ground surface. If the surface is subjected to periodic temperature variations of $T_0 + A\cos\omega t$ over a long time, what is the temperature variation experienced by the egg clutch? (Hint: find a complex, plane wave solution of the diffusion equation of the form $u = T_0 + Ae^{i(\gamma x - \omega t)}$ and determine γ in terms of ω and the diffusivity k of the soil; take the real part). At depth x_1 find the phase shift of the temperature and the amplitude attenuation factor. Plot of the amplitude of the temperature variation versus depth. Take $T_0 = 30$, $A = 15$ degrees Celsius, $k = 0.004$ cm^2 per second, and $\omega = 2\pi$ per day, and plot the temperature variations at the surface and at the depth 3 cm.

2.5 Sources and Duhamel's Principle

How do we proceed if the PDE contains a source term? For example, consider the heat-flow problem

$$u_t = ku_{xx} + f(x, t), \quad x \in \mathbf{R}, t > 0, \tag{2.30}$$

$$u(x, 0) = 0, \quad x \in \mathbf{R}, \tag{2.31}$$

where f is a given heat source. The key to the analysis of this problem can be discovered by examining an ordinary differential equation with a source term. For example, consider the initial value problem

$$y'(t) + ay = F(t), \quad t > 0; \quad y(0) = 0. \tag{2.32}$$

Multiplying by the integrating factor e^{at} makes the left side a total derivative, and we obtain

$$\frac{d}{dt}(e^{at}y) = e^{at}F(t).$$

Integrating from 0 to t then gives

$$e^{at}y(t) = \int_0^t e^{a\tau}F(\tau)d\tau,$$

which can be rewritten as

$$y(t) = \int_0^t e^{-a(t-\tau)}F(\tau)d\tau. \tag{2.33}$$

Now let us consider another problem, where we put the source term in as the initial condition. Let $w = w(t;\tau)$ be the solution to the problem

$$w'(t;\tau) + aw(t;\tau) = 0, \quad t > 0; \quad w(0;\tau) = F(\tau),$$

where a new parameter τ has been introduced. It is straightforward to see that the solution to this problem is

$$w(t;\tau) = F(\tau)e^{-at}.$$

So the solution to (2.32), the problem with a source, is the integral of the solution $w(t,\tau)$ (with t replaced by $t-\tau$) of the associated homogeneous problem where the source is included as an initial condition.

The fact that a particular solution of a linear equation can be deduced from the solution of the homogeneous equation is called *Duhamel's principle*. For ODEs we state the principle as follows:

Duhamel's principle for ODEs
The solution of the problem

$$y'(t) + ay = F(t), \quad t > 0; \quad y(0) = 0$$

is given by

$$y(t) = \int_0^t w(t-\tau,\tau)d\tau,$$

where $w = w(t,\tau)$ solves the homogeneous problem

$$w'(t;\tau) + aw(t;\tau) = 0, \quad t > 0; \quad w(0;\tau) = F(\tau).$$

The same type of result is true for second-order ODEs as well; the reader may recall that the variation of parameters method uses the homogeneous solutions to construct a particular solution. See the Appendix for the formula.

Now let us extrapolate this idea and apply it to the heat flow problem (2.30)–(2.31). If Duhamel's principle is valid in this case, then the solution of (2.30)–(2.31) should be

$$u(x, t) = \int_0^t w(x, t - \tau, \tau) d\tau,$$

where $w(x, t; \tau)$ solves the homogeneous problem

$$w_t = k w_{xx}, \quad x \in \mathbf{R}, t > 0, \tag{2.34}$$

$$w(x, 0; \tau) = f(x, \tau), \quad x \in \mathbf{R}. \tag{2.35}$$

In fact, we can write down the explicit formula; by (2.8) the solution to (2.34)–(2.35) is

$$w(x, t; \tau) = \int_{-\infty}^{\infty} G(x - y, t) f(y, \tau) dy,$$

where G is the heat kernel. Therefore, the solution to (2.30)–(2.31) should be given by

$$u(x, t) = \int_0^t \int_{-\infty}^{\infty} G(x - y, t - \tau) f(y, \tau) dy d\tau.$$

Indeed, one can verify that this is the case. Observe again that τ is being regarded as a parameter in the discussion above.

It is not too surprising that the solution turned out to be an integral. The PDE (2.30) has the form $Hu = f$, where $H = \frac{\partial}{\partial t} - k\frac{\partial^2}{\partial t^2}$ is a differential operator. If we formally write $u = H^{-1} f$ (as we might do in matrix theory if H were a matrix and u and f vectors), then we might expect H^{-1}, the inverse of H, to be an integral operator, since integration and differentiation are inverse processes.

We may now write down the formula for the solution to the problem

$$u_t = k u_{xx} + f(x, t), \quad x \in \mathbf{R}, t > 0, \tag{2.36}$$

$$u(x, 0) = \phi(x), \quad x \in \mathbf{R}, \tag{2.37}$$

where the initial condition is no longer zero. By linearity, the solution to (2.36)–(2.37) is the sum of the solutions to the two problems

$$u_t = k u_{xx}, \quad x \in \mathbf{R}, t > 0, \tag{2.38}$$

$$u(x, 0) = \phi(x), \quad x \in \mathbf{R}, \tag{2.39}$$

and

$$u_t = k u_{xx} + f(x, t), \quad x \in \mathbf{R}, t > 0, \tag{2.40}$$

$$u(x, 0) = 0, \quad x \in \mathbf{R}. \tag{2.41}$$

Thus, the solution to (2.36)–(2.37) is

$$u(x, t) = \int_{-\infty}^{\infty} G(x - y, t)\phi(y) dy + \int_0^t \int_{-\infty}^{\infty} G(x - y, t - \tau) f(y, \tau) dy d\tau. \tag{2.42}$$

This is the *variation of parameters formula* for the problem (2.36)–(2.37), which is the analogue of (2.33) for the ODE (2.32).

Duhamel's principle can also be applied to the wave equation. The solution to the problem

$$u_{tt} - c^2 u_{xx} = f(x,t), \quad x \in \mathbf{R}, t > 0, \tag{2.43}$$

$$u(x,0) = u_t(x,0) = 0, \quad x \in \mathbf{R}, \tag{2.44}$$

is

$$u(x,t) = \int_0^t w(x, t-\tau, \tau)d\tau,$$

where $w = w(x, t; \tau)$ is the solution to

$$w_{tt} - c^2 w_{xx} = 0, \quad x \in \mathbf{R}, t > 0,$$

$$w(x, 0; \tau) = 0, \quad w_t(x, 0; \tau) = f(x, \tau), \quad x \in \mathbf{R}.$$

We put the source term f into the initial condition for w_t rather than for w because f, like w_t, is an acceleration; note that the displacement u is a time integral of w, so w must be a velocity, making w_t an acceleration. ¿From d'Alembert's formula,

$$w(x,t;\tau) = \frac{1}{2c}\int_{x-ct}^{x+ct} f(s,\tau)ds,$$

and therefore by Duhamel's principle the solution to (2.43)–(2.44) is given by the formula

$$u(x,t) = \frac{1}{2c}\int_0^t \int_{x-c(t-\tau)}^{x+c(t-\tau)} f(s,\tau)dsd\tau. \tag{2.45}$$

Source terms also arise in PDEs when problems are transformed in order to homogenize the boundary conditions. For example, consider the diffusion problem

$$u_t = ku_{xx}, \quad x > 0, t > 0, \tag{2.46}$$

$$u(x,0) = \phi(x), \quad x > 0, \tag{2.47}$$

$$u(0,t) = g(t), \quad t > 0. \tag{2.48}$$

We solved this problem in the last section when $g(t) = 0$. So let us attempt to transform the problem into one where the boundary condition is zero. To this end, let $v(x,t) = u(x,t) - g(t)$, or $u(x,t) = v(x,t) + g(t)$; then substituting into (2.46)–(2.48) gives

$$v_t = kv_{xx} - g'(t), \quad x > 0, t > 0, \tag{2.49}$$

$$v(x,0) = \psi(x) \equiv \phi(x) - g(0), \quad x > 0, \tag{2.50}$$

$$v(0,t) = 0, \quad t > 0. \tag{2.51}$$

Therefore, transformation of the dependent variable has changed the problem into one with a homogeneous boundary condition, but a price was paid—an inhomogeneity, or source term $-g'(t)$, was introduced into the PDE. In general, we can always homogenize the boundary conditions in a linear problem, but the result is an inhomogeneous PDE; so inhomogeneous boundary conditions can be traded for inhomogeneous PDEs.

We can solve (2.49)–(2.51) for $v(x, t)$ by formulating a Duhamel's principle. However, in Section 2.5 we will observe that Laplace transform methods can also be applied to find the solution.

Exercises

1. Write a formula for the solution to the problem

$$u_{tt} - c^2 u_{xx} = \sin x, \quad x \in \mathbf{R}, t > 0,$$

$$u(x, 0) = u_t(x, 0) = 0, \quad x \in \mathbf{R}.$$

Graph the solution surface when $c = 1$.

2. Write a formula for the solution to the problem

$$u_t - k u_{xx} = \sin x, \quad x \in \mathbf{R}, t > 0,$$

$$u(x, 0) = 0, \quad x \in \mathbf{R}.$$

3. Using Duhamel's principle, find a formula for the solution to the initial value problem for the convection equation

$$u_t + c u_x = f(x, t), \quad x \in \mathbf{R}, \ t > 0; \qquad u(x, 0) = 0, \quad x \in \mathbf{R}.$$

Hint: Look at the problem

$$w_t(x, t; \tau) + c w_x(x, t; \tau) = 0, \quad x \in \mathbf{R}, t > 0; \qquad w(x, 0; \tau) = f(x, \tau), \quad x \in \mathbf{R}.$$

Solve the problem

$$u_t + 2u_x = xe^{-t}, \quad x \in \mathbf{R}, t > 0; \quad u(x, 0) = 0, \quad x \in \mathbf{R}.$$

4. Formulate Duhamel's principle and solve the initial boundary value problem

$$u_t = k u_{xx} + f(x, t), \quad x > 0, t > 0,$$

$$u(x, 0) = 0, \quad x > 0,$$

$$u(0, t) = 0, \quad t > 0.$$

The solution is

$$u(x, t) = \int_0^t \int_0^\infty (G(x - y, t - \tau) - G(x + y, t - \tau)) f(y, \tau) \, dy \, d\tau.$$

2.6 Laplace Transforms

Laplace transforms are first encountered in elementary differential equations courses as a technique for solving linear, constant-coefficient ordinary differential equations; Laplace transforms convert an ODE into an algebra problem. The ideas easily extend to PDEs, where the the operation of Laplace transformation converts PDEs to ODEs.

Let $u = u(t)$ be a piecewise continuous function on $t \geq 0$ that does not grow too fast; for example, assume that u is of *exponential order*, which means that $|u(t)| \leq c\exp(at)$ for t sufficiently large, where $a, c > 0$. Then the *Laplace transform* of u is defined by

$$(\mathcal{L}u)(s) \equiv U(s) = \int_0^\infty u(t)e^{-st}dt. \tag{2.52}$$

The Laplace transform is an example of an integral transform; it takes a given function $u(t)$ in the time domain and converts it to a new function $U(s)$, in the so-called transform domain. U and s are called the transform variables. The transform is linear in that $\mathcal{L}(c_1 u + c_2 v) = c_1\mathcal{L}u + c_2\mathcal{L}v$, where c_1 and c_2 are constants. If the transform $U(s)$ is known, then $u(t)$ is called the inverse transform of $U(s)$ and we write $\mathcal{L}^{-1}U = u$. Pairs of Laplace transforms and their inverses are tabulated in books of tables, and many software packages contain commands that yield transform pairs. A short table is given at the end of this text. In the Maple computer algebra system, the command

```
laplace (sin(t),t,s);
```

returns

$$\frac{1}{1+s^2},$$

which is the Laplace transform of $\sin t$. To obtain the inverse transform use

```
invlaplace (1/(1+s^2),s,t);
```

which returns

$$\sin t.$$

The importance of the Laplace transform, like other transforms, is that it changes derivative operations to multiplication operations in the transform domain. In fact, we have

$$(\mathcal{L}u')(s) = sU(s) - u(0), \tag{2.53}$$

$$(\mathcal{L}u'')(s) = s^2U(s) - su(0) - u'(0). \tag{2.54}$$

Formulae (2.53) and (2.54) are readily derived using integration by parts, and they are the basic operational formulae for solving differential equations.

Example

Solve the initial value problem

$$u'' + u = 0, \quad t > 0; \quad u(0) = 0, \ u'(0) = 1.$$

Taking Laplace transforms of both sides of the differential equation and using (2.54) gives

$$s^2 U(s) - 1 + U(s) = 0,$$

or

$$U(s) = \frac{1}{1 + s^2}.$$

So, the ordinary differential equation has been transformed into an algebraic equation, and we have solved the problem in the transform domain. To recover $u(t)$ from its transform, we can look up the inverse transform in the table of Laplace transforms or use a computer algebra program to find

$$u(t) = \sin t,$$

which is the solution. □

In summary, taking the Laplace transform of an ordinary differential equation for $u(t)$ results in an algebraic equation for $U(s)$ in the transform domain. We solve the algebraic equation for $U(s)$ and then recover $u(t)$ by inversion. The same strategy applies to partial differential equations where the unknown is a function of two variables, for example $u = u(x, t)$. Now we transform on t, as before, with the variable x being a parameter unaffected by the transform. In particular, we define the Laplace transform of $u(x, t)$ by

$$(\mathcal{L}u)(x, s) \equiv U(x, s) = \int_0^\infty u(x, t)e^{-st}\,dt. \tag{2.55}$$

Then time derivatives transform as in (2.53) and (2.54); for example,

$$(\mathcal{L}u_t)(x, s) = sU(x, s) - u(x, 0).$$

On the other hand, spatial derivatives are left unaffected; for example,

$$(\mathcal{L}u_x)(x, s) = \int_0^\infty \frac{\partial}{\partial x} u(x, t)e^{-st}\,dt = \frac{\partial}{\partial x}\int_0^\infty u(x, t)e^{-st}\,dt = U_x(x, s).$$

Therefore, upon taking Laplace transforms, a PDE in x and t is reduced to an ordinary differential equation in x; all the t-derivatives are turned into multiplication in the transform domain.

Example

Use Laplace transforms to solve the following initial boundary value problem associated with the diffusion equation. Let $u = u(x, t)$ denote the concentration of a chemical contaminant dissolved in a fluid in the semi-infinite domain $x > 0$. Initially, assume that the domain is free from contamination. For times $t > 0$ we impose a constant unit concentration of a contaminant on the boundary $x = 0$, and we ask how this contaminant diffuses into the region. Assuming a unit diffusion constant, the mathematical model is

$$u_t - u_{xx} = 0, \quad x > 0, \ t > 0,$$

$$u(x, 0) = 0, \quad x > 0,$$

$$u(0, t) = 1, \quad t > 0; \quad u(x, t) \text{ bounded.}$$

Taking Laplace transforms of both sides of the PDE gives

$$sU(x, s) - U_{xx}(x, s) = 0.$$

This is an ordinary differential equation with x as the independent variable, and the solution is

$$U(x, s) = a(s)e^{-\sqrt{s}x} + b(s)e^{\sqrt{s}x}.$$

Because we want bounded solutions, we must set $b(s) = 0$. Then

$$U(x, s) = a(s)e^{-\sqrt{s}x}.$$

Now we take a Laplace transform of the boundary condition to get $U(0, s) = 1/s$, where we have used $\mathcal{L}(1) = 1/s$. Therefore, $a(s) = 1/s$, and the solution in the transform domain is

$$U(x, s) = \frac{1}{s}e^{-\sqrt{s}x}.$$

Consulting the table or a computer algebra program, we find that the solution is

$$u(x, t) = \operatorname{erfc}\left(\frac{x}{\sqrt{4t}}\right),$$

where erfc is the function defined by the formula

$$\operatorname{erfc}(y) = 1 - \frac{2}{\sqrt{\pi}}\int_0^y e^{-r^2}\,dr.$$

Observe that $\operatorname{erfc}(y) = 1 - \operatorname{erf}(y)$. □

In both preceding examples we were able recover a function from its Laplace transform by looking up the transform pair in a table or finding it in a software package. One might ask, in general, how to determine $u(t)$ from knowledge of its transform $U(s)$. It would take us far afield to give a thorough discussion. However, we can indicate the general formula used to compute the inverse transform, that is, to compute $u(t)$ from its transform $U(s)$. The *inversion formula* is

$$u(t) = (\mathcal{L}^{-1}U)(t) = \frac{1}{2\pi i}\int_{a-i\infty}^{a+i\infty} U(s)e^{st}\,ds.$$

The integral in this formula is a complex contour integral taken over the infinite straight line (called a Bromwich path) in the complex plane from $a-i\infty$ to $a+i\infty$. The number a is any real number for which the resulting Bromwich path lies to the right of any singularities (poles, essential singular points, or branch points and cuts) of the function $U(s)$. A thorough discussion of the inversion formula can be found in the references.

Another result that is useful in calculations is the convolution theorem.

Theorem

Let u and v be piecewise continuous on $t \geq 0$ and of exponential order. Then

$$\mathcal{L}(u * v)(s) = U(s)V(s),$$

where

$$(u * v)(t) \equiv \int_0^t u(t-\tau)v(\tau)d\tau$$

is the convolution of u and v, and $U = \mathcal{L}u$, $V = \mathcal{L}v$.

We note that the Laplace transform is additive, but it is not multiplicative. That is, the Laplace transform of a product is not the product of the Laplace transforms. The convolution theorem tells what to take the transform of in order to get a product of Laplace transforms, namely, the convolution.

EXAMPLE

In the contaminant transport model in the last example let us change the boundary condition to a function of time and consider

$$u_t - u_{xx} = 0, \quad x > 0,\ t > 0,$$
$$u(x,0) = 0, \quad x > 0,$$
$$u(0,t) = f(t), \quad t > 0; \quad u(x,t) \text{ bounded}.$$

Taking Laplace transforms of both sides of the PDE gives, as in the example,

$$sU(x,s) - U_{xx}(x,s) = 0,$$

which has solution

$$U(x,s) = a(s)e^{-\sqrt{s}x}.$$

Now we take a Laplace transform of the boundary condition to get $U(0, s) = F(s)$. Therefore, $a(s) = F(s)$ and the solution in the transform domain is

$$U(x, s) = F(s)e^{-\sqrt{s}x}.$$

Consulting a table or Maple, we find that

$$\mathcal{L}^{-1}\left(e^{-\sqrt{s}x}\right) = \frac{x}{\sqrt{4\pi t^3}}e^{-x^2/4t}.$$

Therefore, we can use the convolution theorem to write the solution as

$$u(x, t) = \int_0^t \frac{x}{\sqrt{4\pi(t-\tau)^3}}e^{-x^2/(4(t-\tau))}f(\tau)d\tau.$$

□

Contained in the Maple software program there is a package (the integral transform package) that can be adapted to solve PDEs using Laplace transforms. We discuss this procedure in Section 2.8.

A very complete reference for the Laplace transform that includes an extensive table, theory, and applications is Churchill (1958). A short table appears in the back of this text.

Exercises

1. Solve the following problem using Laplace transforms.

$$u_{tt} = c^2 u_{xx} - g, \quad x > 0, \ t > 0,$$

$$u(0, t) = 0, \quad t > 0,$$

$$u(x, 0) = u_t(x, 0) = 0, \quad x > 0.$$

The solution shows what happens to a falling cable lying on a table that is suddenly removed. Draw some time snapshots of the solution.

2. In the quarter plane $x, y > 0$, where the temperature is initially zero, heat flows only in the y-direction; along the edge $y = 0$ heat is convected along the x-axis, and the temperature is constantly 1 at the point $x = y = 0$. The boundary value problem for the temperature $u(x, y, t)$ is

$$u_t = u_{yy}, \quad x, t, y > 0,$$

$$u(x, y, 0) = 0, \quad x, y > 0,$$

$$u(0, 0, t) = 1, \quad t > 0,$$

$$u_t(x, 0, t) + u_x(x, 0, t) = 0, \quad x, t > 0.$$

Find a bounded solution to this problem using Laplace transforms.

3. Show that $\mathcal{L}\left(\int_0^t f(\tau)d\tau\right) = F(s)/s$.

4. Show that $\mathcal{L}\left(H(t-a)f(t-a)\right) = e^{-as}F(s)$, where H is the unit step function defined by $H(x) = 0$ for $x < 0$, and $H(x) = 1$ for $x \geq 0$.

5. A very deep container of liquid is insulated on its sides. Initially, its temperature is a constant u_0 degrees, and for $t > 0$ heat radiates from its exposed top surface according to Newton's law of cooling (see Section 1.3). The air temperature is zero degrees. Formulate an initial boundary value problem for the temperature of the liquid, and find a formula for the temperature at various depths at various times. Use $\rho = c = K = \beta = 1$ and use a computer algebra package to graph some temperature profiles.

6. Derive the solution $u(x,t) = H(t - x/c)g(t - x/c)$ to the problem

$$u_{tt} = c^2 u_{xx}, \quad x, t > 0,$$

$$u(x,0) = u_t(x,0) = 0, \quad x > 0,$$

$$u(0,t) = g(t), \quad t > 0.$$

2.7 Fourier Transforms

The Fourier transform is another integral operator with properties similar to the Laplace transform in that derivatives are turned into multiplication operations in the transform domain. Thus the Fourier transform, like the Laplace transform, is useful as a computational tool in solving differential equations. In PDEs the Laplace transform is usually applied to the time variable; the Fourier transform is usually applied to the spatial variable when it varies over $(-\infty, \infty)$.

First let us begin with functions of one variable. The Fourier transform of a function $u = u(x)$, $x \in \mathbf{R}$, is defined by the equation

$$(\mathcal{F}u)(\xi) \equiv \hat{u}(\xi) = \int_{-\infty}^{\infty} u(x)e^{i\xi x}\,dx. \tag{2.56}$$

If u is absolutely integrable, i.e., $\int_{-\infty}^{\infty} |u|dx < \infty$, then $\hat{u}$ can be shown to exist. However, in the theory of Fourier transforms, it is common to work with a a smaller set of functions; the integrals involved are improper integrals, and so the functions must decay rapidly enough in order for the integrals to exist. We define $\mathcal{S}$ as the set of rapidly decreasing functions on $\mathbf{R}$ that have continuous derivatives of all orders; the rapidly decreasing functions are functions that, along with all their derivatives, decay to zero as $x \to \pm\infty$ faster than any power function (functions like $1/x^2$ and $1/x^6$). The function $\exp(-x^2)$ defining the bell-shaped curve is a such a rapidly decreasing function. More technically, if the set of functions that have continuous derivatives of all orders on $\mathbf{R}$ is denoted by C^∞, then

$$\mathcal{S} = \{u \in C^\infty : |\frac{d^k u}{dx^k}| \le M \frac{1}{|x|^N} \text{ as } |x| \to \infty,$$

$$k = 0, 1, 2, \ldots; \text{ for all integers } N\}.$$

The set $\mathcal{S}$ is called the *Schwartz class* of functions, and one can show that if $u \in \mathcal{S}$, then $\hat{u} \in \mathcal{S}$, and conversely. So $\mathcal{S}$ is a closed set under Fourier transformation, which makes it a good set to work with.

There is one important remark about notation. There is no standard convention on how to define the Fourier transform (2.56); some put a factor of $1/(2\pi)$ or $1/\sqrt{2\pi}$ in front of the integral, and some have a negative power in the exponential, or even a factor of 2π. One should be aware of these variations when consulting other sources.

A basic property of the Fourier transform is that the kth derivative $u^{(k)}$ ($k = 1, 2, \ldots$) transforms to an algebraic expression. That is,

$$(\mathcal{F}u^{(k)})(\xi) = (-i\xi)^k \hat{u}(\xi), \quad u \in \mathcal{S}, \tag{2.57}$$

confirming our comment that derivatives are transformed to multiplication (by a factor of $(-i\xi)^k$). This formula is easily proved by integration by parts (as for the Laplace transform); all the boundary terms generated in the integration by parts are zero, since u and all its spatial derivatives vanish at $\pm\infty$. For functions of two variables, say $u = u(x, t)$, the variable t acts as a parameter, and we define

$$(\mathcal{F}u)(\xi, t) \equiv \hat{u}(\xi, t) = \int_{-\infty}^{\infty} u(x, t)e^{i\xi x}dx.$$

Then, under Fourier transformation, x-derivatives turn into multiplication, and t derivatives remain unaffected; for example,

$$(\mathcal{F}u_x)(\xi, t) = (-i\xi)\hat{u}(\xi, t),$$

$$(\mathcal{F}u_{xx})(\xi, t) = (-i\xi)^2\hat{u}(\xi, t),$$

$$(\mathcal{F}u_t)(\xi, t) = \hat{u}_t(\xi, t).$$

Solving a differential equation for u involves first transforming the problem into the transform domain and then solving for $\hat{u}$. Then one is faced with the inversion problem, or the problem of determining the u for which $\mathcal{F}u = \hat{u}$. Another nice property of the Fourier transform is the simple form of the inversion formula, or inverse transform. It is

$$(\mathcal{F}^{-1}\hat{u})(x) \equiv u(x) = \frac{1}{2\pi}\int_{-\infty}^{\infty} \hat{u}(\xi)e^{-i\xi x}d\xi. \tag{2.58}$$

This result is called the *Fourier integral theorem*; it dictates how to get back from the transform domain.

Some Fourier transforms can be calculated directly; many require complex contour integration. In the following example we calculate the transform of the Gaussian function $u(x) = e^{-ax^2}$, $a > 0$, using a differential equation technique.

EXAMPLE
We want to calculate $\hat{u}$, where

$$\hat{u}(\xi) = \int_{-\infty}^{\infty} e^{-ax^2} e^{i\xi x} dx.$$

Differentiating with respect to ξ and then integrating by parts gives

$$\begin{aligned}\hat{u}'(\xi) &= i \int_{-\infty}^{\infty} x e^{-ax^2} e^{i\xi x} dx \\ &= \frac{-i}{2a} \int_{-\infty}^{\infty} \frac{d}{dx} e^{-ax^2} e^{i\xi x} dx \\ &= \frac{-\xi}{2a} \hat{u}(\xi).\end{aligned}$$

Therefore, we have a differential equation $\hat{u}' = \frac{-\xi}{2a}\hat{u}$ for $\hat{u}$. Separating variables and integrating gives the general solution

$$\hat{u}(\xi) = Ce^{-\xi^2/(4a)}.$$

The constant C can be determined by noticing that

$$\hat{u}(0) = \int_{-\infty}^{\infty} e^{-ax^2} dx = \sqrt{\frac{\pi}{a}}.$$

Consequently, we have

$$\mathcal{F}(e^{-ax^2}) = \sqrt{\frac{\pi}{a}} e^{-\xi^2/(4a)}. \tag{2.59}$$

So, the Fourier transform of a Gaussian function is itself a Gaussian; likewise, the inverse transform of a Gaussian is a Gaussian. □

A convolution relation also holds for Fourier transforms. If $u, v \in \mathcal{S}$, then we define their *convolution*, which is in $\mathcal{S}$, by

$$(u * v)(x) = \int_{-\infty}^{\infty} u(x - y)v(y)dy.$$

Then we have the following *convolution theorem*:

Theorem
If $u,v \in \mathcal{S}$, then

$$\mathcal{F}(u * v)(\xi) = \hat{u}(\xi)\hat{v}(\xi).$$

By the Fourier integral theorem it follows immediately that

$$(u * v)(x) = \mathcal{F}^{-1}(\hat{u}(\xi)\hat{v}(\xi)).$$

This formula states that the inverse transform of a product is a convolution; this is a useful relationship in solving differential equations.

Example
Let $f \in \mathcal{S}$ and determine u for which

$$u'' - u = f(x), \quad x \in \mathbf{R}.$$

Taking the transform of both sides yields

$$(-i\xi)^2\hat{u} - \hat{u} = \hat{f},$$

or

$$\hat{u}(\xi) = -\frac{1}{1+\xi^2}\hat{f}(\xi).$$

In the transform domain the solution is a product of transforms, and so we apply the convolution theorem. From Exercises 2 and 3 we have

$$\mathcal{F}\Big(\frac{1}{2}e^{-|x|}\Big) = \frac{1}{1+\xi^2}.$$

Therefore,

$$u(x) = -\frac{1}{2}e^{-|x|} * f(x) = -\frac{1}{2}\int_{-\infty}^{\infty} e^{-|x-y|}f(y)dy. \qquad \square$$

The strategy in applying transform methods to solve differential equations is to proceed formally, making any assumptions that are required to obtain an answer; for example, assume that all the data is in $\mathcal{S}$. When a solution is obtained one can then attempt to verify that it is indeed a solution to the problem. Often one can prove that the solution obtained holds under less severe conditions than are required in application of the transform method.

Now we apply the Fourier transform method to the Cauchy problem for the heat equation. We will derive the same solution formula (2.8) that we obtained in Section 2.1 by a different method.

Example
Use Fourier transforms to solve the pure initial value problem for the heat equation:

$$u_t - ku_{xx} = 0, \quad x \in \mathbf{R},\ t > 0; \quad u(x,0) = f(x),\ x \in \mathbf{R}. \tag{2.60}$$

Again we assume that $f \in \mathcal{S}$. Taking Fourier transforms of the PDE gives

$$\hat{u}_t = -\xi^2 k\hat{u},$$

which is an ordinary differential equation in t for $\hat{u}(\xi, t)$, with ξ as a parameter. Its solution is

$$\hat{u}(\xi, t) = Ce^{-\xi^2 kt}.$$

But the initial condition gives $\hat{u}(\xi, 0) = \hat{f}(\xi)$, and so $C = \hat{f}(\xi)$. Therefore,

$$\hat{u}(\xi, t) = e^{-\xi^2 kt}\hat{f}(\xi).$$

Replacing a by $1/(4kt)$ in formula (2.59) gives

$$\mathcal{F}\left(\frac{1}{\sqrt{4\pi kt}}e^{-x^2/(4kt)}\right) = e^{-\xi^2 kt}.$$

Thus, by the convolution theorem we have

$$u(x, t) = \int_{-\infty}^{\infty} \frac{1}{\sqrt{4\pi kt}} e^{-(x-y)^2/(4kt)} f(y)dy. \tag{2.61}$$

This solution was derived under the assumption that f is in the Schwartz class. But now that we have it, we can attempt to show that it is a solution under milder restrictions on f. For example, one can prove that (2.61) is a solution to (2.60) if f is a continuous, bounded function on $\mathbf{R}$. □

EXAMPLE

Now we work a problem involving Laplace's equation in the upper half plane. Consider

$$u_{xx} + u_{yy} = 0, \quad x \in \mathbf{R}, \ y > 0; \quad u(x, 0) = f(x), \quad x \in \mathbf{R}.$$

We also append the condition that the solution u stay bounded as $y \to \infty$. This example is similar to the last example. Taking the transform (on x with y as a parameter) of the PDE, we obtain

$$\hat{u}_{yy} - \xi^2 \hat{u} = 0,$$

which has general solution

$$\hat{u}(\xi, y) = a(\xi)e^{-\xi y} + b(\xi)e^{\xi y}.$$

The boundedness condition on u forces $b(\xi) = 0$ if $\xi > 0$ and $a(\xi) = 0$ if $\xi < 0$. So we take

$$\hat{u}(\xi, y) = c(\xi)e^{-|\xi| y}.$$

Upon applying Fourier transforms to the boundary condition, we get $c(\xi) = \hat{f}(\xi)$. Therefore, the solution in the transform domain is

$$\hat{u}(\xi, y) = e^{-|\xi| y}\hat{f}(\xi).$$

Therefore, using the convolution theorem and Exercise 2, we obtain the solution

$$u(x, y) = \frac{y}{\pi}\frac{1}{x^2 + y^2} * f = \frac{y}{\pi}\int_{-\infty}^{\infty} \frac{f(\tau)d\tau}{(x-\tau)^2 + y^2}.$$

□

An excellent introduction to Fourier transforms can be found in Strichartz (1994).

Exercises

1. Find the convolution of the functions $f(x) = x$ and $g(x) = e^{-x^2}$.

2. Show that the inverse Fourier transform of $e^{-a|\xi|}$, $a > 0$, is

$$\frac{a}{\pi}\frac{1}{x^2 + a^2}.$$

3. Verify the following properties of the Fourier transform:

   (a) $(\mathcal{F}u)(\xi) = 2\pi(\mathcal{F}^{-1}u)(-\xi)$.
   (b) $\mathcal{F}(e^{iax}u)(\xi) = \hat{u}(\xi + a)$.
   (c) $\mathcal{F}(u(x + a)) = e^{-ia\xi}\hat{u}(\xi)$.

   Formula (a) states that if a transform is known, so is its inverse, and conversely.

4. Find the Fourier transform of the function u defined by $u(x) = e^{-ax}$ if $x > 0$, and $u(x) = 0$ if $x \le 0$.

5. Compute $\mathcal{F}(xe^{-ax^2})$.

6. Solve the following initial value problem for the inhomogeneous heat equation:

$$u_t = u_{xx} + F(x, t), \quad x \in \mathbf{R},\ t > 0 \quad u(x, 0) = 0, \quad x \in \mathbf{R}.$$

7. Find a formula for the solution to the following initial value problem for the free Schrödinger equation:

$$u_t = iu_{xx}, \quad x \in \mathbf{R},\ t > 0; \quad u(x, 0) = e^{-x^2}, \ x \in \mathbf{R}.$$

8. Find a bounded solution to the Neumann problem

$$u_{xx} + u_{yy} = 0, \quad x \in \mathbf{R},\ y > 0,$$
$$u_y(x, 0) = g(x), \quad x \in \mathbf{R}.$$

   Hint: Let $v = u_y$ and reduce the problem to a Dirichlet problem. The solution is

$$u(x, y) = \frac{1}{2\pi}\int_{-\infty}^{\infty} g(x - \xi)\ln(y^2 + \xi^2)d\xi + C.$$

9. Solve the boundary value problem

$$u_{xx} + u_{yy} = 0, \quad x \in \mathbf{R},\ y > 0,$$
$$u(x, 0) = 1, \ |x| \le l; \quad u(x, 0) = 0, \quad |x| > l,$$

   where $b > 0$ is a constant.

10. Use integration by parts to verify (assume $u \in \mathcal{S}$)

$$(\mathcal{F}u_x)(\xi, t) = (-i\xi)\hat{u}(\xi, t),$$
$$(\mathcal{F}u_{xx})(\xi, t) = (-i\xi)^2\hat{u}(\xi, t).$$

11. Let $u(x)$ be a square wave, i.e, $u(x) = 1$ if $|x| \leq a$ and $u(x) = 0$ if $|x| > a$. Show that

$$(\mathcal{F}u)(\xi) = \frac{2 \sin a\xi}{\xi}.$$

12. Solve the Cauchy problem for the following convection–diffusion equation using Fourier transforms:

$$u_t = Du_{xx} - cu_x, \quad x \in \mathbf{R}, \ t > 0; \quad u(x, 0) = \phi(x), \quad x \in \mathbf{R}.$$

13. This exercise explores the role of a term u_{xxx} (called a *dispersion* term) in a PDE by examining the equation

$$u_t + u_{xxx} = 0.$$

This equation is sometimes called the linearized Korteweg–deVries (KdV) equation.

(a) What relation between ω and k would have to hold if a solution of the form

$$u(x, t) = e^{i(kx-\omega t)}$$

exists? What do these solutions look like, and how does their speed depend on k? What does your conclusion mean, qualitatively? (Recall that the real and imaginary parts of a complex solution to a linear equation are both real solutions.)

(b) Use Fourier transforms to solve the Cauchy problem for the linearized KdV equation, and write your answer in terms of the *Airy function* defined by

$$\text{Ai}(x) = \frac{1}{2\pi} \int_{-\infty}^{\infty} \cos(z^3/3 + xz)dz.$$

(c) Take a Gaussian ($\exp(-x^2)$) as the initial condition in the Cauchy problem and find the solution profile when $t = 1$. (Both Maple and Mathematica know the Airy function.)

14. Use Fourier transforms to derive d'Alembert's solution (2.14) to the Cauchy problem for the wave equation (2.12)–(2.13) when the initial velocity is zero, i.e., $g(x) \equiv 0$.

2.8 Solving PDEs Using Computer Algebra Systems

Many computer algebra systems (CAS) contain simple commands that return the general solution of certain classes of PDEs or a graphical solution to a Cauchy-type problem. In this section we illustrate some of these commands with Maple. Similar commands are a part of Mathematica. There are many manuals that describe these computer algebra systems. The reader should recognize that the entire field of symbolic calculations for PDEs is advancing at a rapid pace, and therefore the material in this section may become outdated quickly as newer versions of these programs incorporate the most recent developments. The practitioner must be careful when using software packages; the lesson is that one cannot proceed blindly without applying some basic knowledge of PDEs.

The basic command in Maple is `pdsolve`, which has the syntax

`pdsolve(`*PDE, var*`)`

where *PDE* is the partial differential equation and *var* is the dependent variable. The `pdsolve` program returns the general solution, in terms of arbitrary functions, for a large number of partial differential equations. For example, to define the wave equation, at the command prompt we type

```
PDE:=diff(u(x,t),t,t)-c^2*diff(u(x,t),x,x)=0;
```

Maple returns a printed statement of the PDE. Then to solve it, we type

```
sol:=pdsolve (PDE,u(x,t));
```

and Maple returns the general solution

```
sol:=u(x,t)=_F1(tc+x)+_F2(tc-x)
```

where `_F1` and `_F2` are Maple's notation for two arbitrary functions.

If we want the general solution, which is in terms of arbitrary functions, then a CAS can be extremely time-saving. But the reader should bear in mind that we are not always interested in the general solution to a PDE. More often, we seek a particular solution that satisfies initial and boundary conditions. In this case the general solution may be of help if we can evaluate the arbitrary functions using the auxiliary conditions; but often this strategy fails. Moreover, many PDEs (e.g., the heat equation) do not have a general solution that has a concise form like that of the wave equation and `pdsolve` cannot solve these types of equations.

Knowledge of the general solution to a first-order PDE is helpful because we use the boundary or initial data to specify the arbitrary function that appears in the general solution.

Example
Consider

$$u_t + 2u_x = -3u, \quad x \in \mathbf{R}, t > 0; \quad u(x,0) = \frac{1}{1+x^2}, \quad x \in \mathbf{R}.$$

This equation models the advection and decay of a chemical substance in an infinite domain. The general solution is

$$u(x,t) = g(2t - x)e^{-3x/2},$$

which is found from the Maple command

```
pdsolve(diff(u(x,t),t)+2*diff(u(x,t),x)
= - 3*u(x,t);
```

Now we use the initial data to determine the arbitrary function g. Of course, we can use the CAS to perform this calculation for us, but it is often simpler to just do it by hand. Setting $t = 0$ in the general solution gives

$$g(-x)e^{-3x/3} = \frac{1}{1+x^2}.$$

Thus

$$g(x) = e^{-3x/3}\frac{1}{1+x^2}$$

and the solution is

$$u(x,t) = e^{-3(2t-x)/3}\frac{1}{1+(2t-x)^2}e^{-3x/2} = \frac{e^{-3t}}{1+4t^2-4tx+x^2}.$$

This solution can be plotted on $-3 < x < 3,\ 0 < t < 1.5$ using Maple's plot command

```
plot3d(u(x,t),x=-3..3,t=0..1.5);
```

The solution surface is shown in Figure 2.7. □

Using a computer algebra system like Maple is not the only way to use computers to solve PDEs. The reader should recall from an elementary course in ODEs that there are discrete methods to numerically solve ordinary differential equations (Euler's method, the Runge–Kutta method, etc.). Similar methods, based on difference-quotient approximations of the partial derivatives, can be developed for partial differential equations. We shall introduce some of these methods in Section 4.8. At present, it can be stated that it is more common in real science and engineering

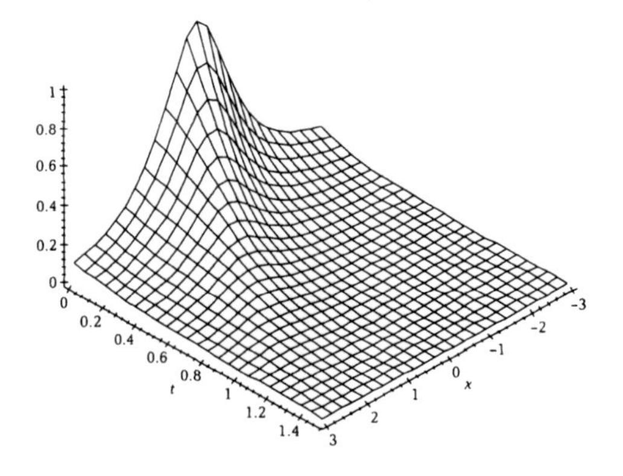

Figure 2.7. The solution surface $u(x,t) = e^{-3t}/(1+4t^2-4tx+x^2)$.

problems to solve PDEs using numerical methods (finite differences, finite elements, etc.) rather than use exact methods from a CAS. Highly efficient numcrical algorithms can be written in FORTRAN, C++, or Matlab, which is a front end to FORTRAN routines.

Exercises

1. Use pdsolve in Maple, or the appropriate command in another system, to find the general solution of the following PDEs:

   (a) $u_{tt} - 5u = 0$.
   (b) $u_{tt} - u_{xt} - 6u_{xx} = 0$.
   (c) $u_t + \frac{1}{1+x^2} u_x + u = 0$.
   (d) $u_{xt} + u_x = xte^{-t}$.

2. Use pdsolve to find the general solution to the PDE

$$u_t + xu_x = -\sin x$$

   Find the solution that satisfies the boundary condition $u(1, t) = e^{-t^2}$.

3. Use pdsolve to find the general solution to $u_t + u_x = u^2$.

CHAPTER 3 Orthogonal Expansions

3.1 The Fourier Method

The basic technique for solving PDEs on a bounded spatial domain is the Fourier method, named after Joseph Fourier (1768–1830). In this section we take a nineteenth-century perspective and make some comments about the origin of the method. Our discussion will motivate one of the fundamental topics in analysis and in PDEs, namely orthogonal expansions.

To fix the notion let us consider heat flow in a finite bar of length π and unit diffusivity ($k = 1$), where the ends are held at zero degrees. We choose π as the length to make the constants come out simpler at the end. From preceding sections we know that the temperature $u = u(x, t)$ in the bar must satisfy the model equations

$$u_t = u_{xx}, \quad 0 < x < \pi; \quad t > 0, \tag{3.1}$$

$$u(0, t) = u(\pi, t) = 0, \quad t > 0, \tag{3.2}$$

consisting of the heat equation and the boundary conditions. It is easy to check that for any positive integer n a solution to (3.1) and (3.2) is given by

$$u_n(x, t) = e^{-n^2 t} \sin nx.$$

These solutions clearly satisfy the initial conditions

$$u_n(x, 0) = \sin nx,$$

which give temperature profiles at $t = 0$.

But what if the initial temperature profile is not $\sin nx$, but rather an arbitrarily given function $f(x)$? That is, what is the solution to (3.1) and (3.2) augmented by the initial condition

$$u(x, 0) = f(x), \quad 0 < x < \pi? \tag{3.3}$$

Fourier argued that the solution $u(x, t)$ to (3.1)–(3.3) could be taken as an infinite linear combination of the $u_n(x, t)$, that is,

$$u(x, t) = \sum_{n=1}^{\infty} a_n e^{-n^2 t} \sin nx \tag{3.4}$$

for appropriately chosen constants a_n. To satisfy the initial condition, (3.3) would require that

$$u(x, 0) = f(x) = \sum_{n=1}^{\infty} a_n \sin nx, \tag{3.5}$$

that is, the initial temperature function $f(x)$ would have to have a representation in terms of the periodic functions $\sin x$, $\sin 2x$, $\sin 3x$, ... To determine such a_n one could proceed formally and multiply both sides of (3.5) by $\sin mx$, for some fixed but arbitrary index m, and then integrate from $x = 0$ to $x = \pi$ to obtain

$$\int_0^{\pi} f(x) \sin mx \, dx = \int_0^{\pi} \sum_{n=1}^{\infty} a_n \sin nx \sin mx \, dx.$$

Now if we interchange the order of integration (we alert the reader that this is not always a valid operation), we obtain

$$\int_0^{\pi} f(x) \sin mx \, dx = \sum_{n=1}^{\infty} a_n \int_0^{\pi} \sin nx \sin mx \, dx.$$

Now we observe that the infinite series collapses to a single term. This is because of the integration formula

$$\int_0^{\pi} \sin nx \sin mx \, dx = 0, \quad n \neq m. \tag{3.6}$$

So the only term that survives in the infinite series is the one where the summation index n hits m, the fixed index. Then we get

$$\int_0^{\pi} f(x) \sin mx \, dx = a_m \int_0^{\pi} \sin^2 mx \, dx = \frac{\pi}{2} a_m.$$

But m is an arbitrary index, and so we have shown that the coefficients in (3.4) are given by

$$a_n = \frac{2}{\pi} \int_0^{\pi} f(x) \sin nx \, dx, \quad n = 1, 2, \ldots. \tag{3.7}$$

So, in a formal sense (a formal calculation in mathematics is one done without complete rigor) the solution to the heat flow problem (3.1)–(3.3) is given by (3.4), where the coefficients a_n are given by (3.7). Indeed, one can verify that this is a solution for reasonable initial temperature distributions $f(x)$.

What we have just described is Fourier's method; namely, we solve a general initial boundary value problem by superimposing the solutions to the pure boundary value problem and then choosing the constants such that the initial condition is satisfied as well. One key to the calculation was that the functions $\sin nx$ satisfied a relation like (3.6), which is called an orthogonality condition. It enabled us to easily calculate the coefficients in the series (3.5). The infinite series (3.5) for $f(x)$ in terms of the orthogonal functions $\sin nx$ is called an orthogonal expansion, or Fourier series. In the sequel we shall examine these ideas in detail.

On another historical note, this calculation was troublesome in Fourier's time, in the early 1800s. Concepts like convergence of series were not well understood. L. Euler, D. Bernoulli, and d'Alembert, all of whom had addressed similar problems in the mid-1700s regarding the wave equation and vibrating strings, had wondered about the possibility of expanding a nonperiodic function in terms of periodic functions like sines and cosines. It was L. Dirichlet who in 1829 established conditions under which such series representations are valid.

Exercises

1. Consider the initial boundary value problem for the wave equation

$$u_{tt} = c^2 u_{xx}, \quad 0 < x < \pi, \ t > 0, \tag{3.8}$$

$$u(0, t) = u(\pi, t) = 0, \quad t > 0, \tag{3.9}$$

$$u(x, 0) = f(x), \quad u_t(x, 0) = 0, \ 0 < x < \pi, \tag{3.10}$$

on a bounded spatial domain. Use the fundamental set of solutions

$$u_n(x, t) = \cos nct \sin nx, \quad n = 1, 2, \ldots,$$

which satisfy (3.8) and (3.9), to determine a formal solution of (3.8)–(3.10). Find the solution if the initial conditions are changed to

$$u(x, 0) = 0, \quad u_t(x, 0) = g(x), \ 0 < x < \pi.$$

Observe that these calculations amount to resolving an arbitrary wave into its fundamental modes.

3.2 Orthogonal Expansions

The reader is familiar with the notion of expanding a given function $f(x)$ in an infinite series of simpler functions. For example, if f has infinitely many derivatives at $x = 0$, then its Taylor series about $x = 0$ is

$$f(x) = \sum_{n=0}^{\infty} a_n x^n,$$

where the a_n, the Taylor coefficients, are given by $a_n = f^{(n)}(0)/n!$. For example, the Taylor series for the exponential function is

$$e^x = 1 + x + \frac{1}{2!}x^2 + \frac{1}{3!}x^3 + \cdots.$$

Another type of series is a Fourier series; if f is defined and integrable on the interval $[-\pi, \pi]$, then its *Fourier series* is an infinite series of the form

$$\frac{a_0}{2} + \sum_{n=1}^{\infty}(a_n \cos nx + b_n \sin nx), \tag{3.11}$$

where the coefficients are calculated by the formulae

$$a_n = \frac{1}{\pi}\int_{-\pi}^{\pi} f(x)\cos nx dx, \quad n = 0, 1, 2\ldots,$$

$$b_n = \frac{1}{\pi}\int_{-\pi}^{\pi} f(x)\sin nx dx, \quad n = 1, 2\ldots.$$

In the case of Taylor series we use the set of power functions $1, x, x^2, \ldots$ for the basis of the expansion, and in Fourier series we use the trigonometric functions $1, \cos x, \cos 2x, \ldots, \sin x, \sin 2x, \ldots$ as the basis of the expansion. Each expansion has its advantages. We learn in calculus that a few terms of the Taylor series can provide a good approximation to f locally near $x = 0$. The first few terms of a Fourier series can, on the other hand, provide a good global approximation to periodic signals.

In this section we want to understand the underlying idea of a Fourier expansion by allowing the basis functions to be any set of functions with an *orthogonality* property. The result will be a general theory that incorporates Fourier series as a special case. Orthogonality is a notion that has its origins in vector analysis; we recall that a set of vectors $e_1, e_2, \ldots$ will be mutually perpendicular if their corresponding dot products vanish, i.e., $e_i \cdot e_j = 0$ for all $i \neq j$. Orthogonal is another word for perpendicular, but it has meaning in more general settings, such as with respect to sets of functions. To fix the notion let us define the set of functions $L^2 \equiv L^2[a, b]$ consisting of all the real-valued functions f defined on the interval $[a, b]$ for which

$$\int_a^b |f(x)|^2 dx < \infty.$$

When this property holds, we say that f is *square integrable** and write $f \in L^2[a, b]$. In this set of functions we define a so-called scalar product,

*The integral here is, appropriately, the Lebesgue integral studied in advanced analysis courses rather than the Riemann integral, which is studied in elementary calculus;

which is a generalization of the dot product for vectors. If $f, g \in L^2$, then the *scalar product* of f and g is a real number, denoted by (f, g), and is defined by

$$(f, g) = \int_a^b f(x)g(x)dx.$$

It is easy to see, by the properties of integrals, that the scalar product of functions satisfies the same properties of dot products of vectors:

$$(f, g) = (g, f) \qquad \text{commutativity,}$$

$$(f, g + h) = (f, g) + (f, h) \qquad \text{additivity,}$$

$$(f, \alpha g) = \alpha(f, g), \quad \alpha \in \mathbf{R}, \qquad \text{homogeneity.}$$

We also introduce the idea of length, or size, in the set $L^2[a, b]$. If $f \in L^2[a, b]$, then we define the *norm* of f to be the nonnegative real number

$$\|f\| \equiv \sqrt{(f, f)} = \sqrt{\int_a^b |f(x)|^2 dx}.$$

Observe that this definition is exactly analogous to the definition of the length of a vector in Euclidean space.

EXAMPLE

Let $f(x) = x^3$ and $g(x) = 1 + x$ on the interval $[0, 1]$. Then the scalar product is

$$(f, g) = \int_0^1 x^3(1 + x)dx = \frac{9}{20}.$$

The norm, or size, of f is measured by

$$\|f\| = \sqrt{\int_0^1 (x^3)^2 dx} = \sqrt{\frac{1}{7}}.$$

Notice that the norm is the square root of the area under the graph of the square of the given function. □

We say that a set of functions $f_1, f_2, f_3, \ldots$ in $L^2[a, b]$ is *orthogonal* if $(f_i, f_j) = 0$ for $i \neq j$. For example, the set 1, $\cos x$, $\cos 2x$, ..., $\sin x$, $\sin 2x$, ... is orthogonal on $[-\pi, \pi]$ because if we integrate the product of any two of these over the interval $[-\pi, \pi]$ we get zero. The reader should verify this statement using formulas from a calculus book. For another example, we observed in the last section that

$$\int_0^\pi \sin mx \sin nx dx = 0, \quad n \neq m.$$

however, there will be no harm in interpreting the integrals in this text as a Riemann integrals since all of the functions that we examine will be Riemann integrable (and therefore Lebesgue integrable).

Thus the set of functions $\sin x, \sin 2x, \ldots$ is orthogonal on the interval $[0, \pi]$. The reader should check that the power functions $1, x, x^2, \ldots$ that appear in the Taylor series do *not* form an orthogonal set on any interval $[a, b]$. We remark that functions can be *normalized* in the same way that a vector can be turned into a unit vector, namely by dividing by its size. For functions we divide by the norm. If $f \in L^2[a, b]$, then the function g defined by $g(x) = f(x)/\|f\|$ has unit norm. If each function in an orthogonal set of functions has unit norm, then the set is called *orthonormal*. Thus, orthogonal sets of functions can always be turned into orthonormal sets.

EXAMPLE

The set of functions

$$\sin \frac{n\pi x}{l}, \quad n = 1, 2, \ldots,$$

is orthogonal on the interval $[0, l]$, and $\| \sin \frac{n\pi x}{l} \|^2 = l/2$. Therefore, the functions

$$\sqrt{\frac{2}{l}} \sin \frac{n\pi x}{l}$$

form an orthonormal set of functions on $[0, l]$. □

The question that we ask is as follows. Can we expand an arbitrary function $f \in L^2$ in a series of given orthogonal functions $f_1, f_2, \ldots$? That is, can we write

$$f(x) = \sum_{n=1}^{\infty} c_n f_n(x) \tag{3.12}$$

for some set of constants c_n, and in what sense can we interpret the expansion? Here, intuitively, we can think of f as a signal and $f_1, f_2, \ldots$ as fundamental modes. So, can we break a signal down into fundamental modes? Notice that this is the same question that we asked specifically in the last section regarding the feasibility of expanding a given initial, arbitrary temperature $f(x)$ in a sine series. By the way, for notational convenience we shall sometimes drop the explicit dependence of functions on x and write just f in place of $f(x)$. In some instances, when f represents a signal, we will think of x as time.

When we write a series like (3.12), a natural question is, "What is meant by convergence of the infinite series?" There are several possible answers to this question, depending on how we measure the error. To formulate some of these answers let us define what we mean by the error in an N-term approximation $s_N(x) \equiv \sum_{n=1}^{N} c_n f_n(x)$ to $f(x)$. On the one hand, by the *pointwise error* $E_N(x)$ we mean the expression

$$E_N(x) \equiv f(x) - s_N(x) \equiv f(x) - \sum_{n=1}^{N} c_n f_n(x).$$

Thus, for a fixed number N (giving the number of terms), $E_N(x)$ is a *function* whose value gives the error at each point x. On the other hand, we can define an *integrated* error; the number e_N defined by

$$e_N \equiv \int_a^b |\, f(x) - s_N(x)\, |^2 \, dx \equiv \int_a^b |\, f(x) - \sum_{n=1}^{N} c_n f_n(x)\, |^2 \, dx$$

is called the *mean-square error* in the approximation. Note that e_N is just the square of the pointwise error, integrated over the interval $[a, b]$. Thus, the mean-square error is a sort of average over all the pointwise errors. Each of these error expressions leads to a definition of convergence.

If for each *fixed* x in $[a, b]$ we have $\lim_{N\to\infty} E_N(x) = 0$, then we say that the infinite series (3.12) converges *pointwise* to f on the interval $[a, b]$; in this case, for each fixed x the series is a numerical series that converges to the numerical value $f(x)$, at that x; pointwise convergence means that the error goes to zero at each point. This is the type of convergence that a calculus student first encounters. We say that the infinite series (3.12) converges to f in the *mean-square* sense if

$$\lim_{N\to\infty} e_N = 0.$$

Mean-square convergence is also called *convergence in* $L^2[a, b]$. It requires that the integrated pointwise error-squared go to zero as more and more terms are taken. In advanced courses it is shown that pointwise convergence does not imply mean-square convergence, and conversely.

There is a stronger definition of convergence, called uniform convergence, that implies both mean-square and pointwise convergence. We say that the series (3.12) converges *uniformly* to $f(x)$ on $[a, b]$ if for any given tolerance $\epsilon > 0$, we can find a number N (representing the number of terms) for which $|E_N(x)| < \epsilon$ *for all* x in $[a, b]$. That is, the error can be made uniformly small over the entire interval by choosing the number of terms large enough. Here, in uniform convergence, N (the number of terms) depends only on the tolerance ϵ and not on where in the interval the error is taken.

The novice may not appreciate the subtle differences in convergence and wonder why we have defined three types (actually, there are others). Suffice it to say that the error may not go to zero in a pointwise or uniform manner, but that does not mean the approximation is not useful; a weaker form of convergence, like mean-square, may be all that is required.

The mean-square convergence criterion can be written simply in terms of the norm of a function. The series (3.12) converges to f in the mean-square sense if and only if $\lim_{N\to\infty} \|E_N\| = 0$. By the way, all of the definitions above are valid whether or not the functions $f_n(x)$ are mutually orthogonal.

Now suppose (3.12) holds in the sense of mean-square convergence where f is a given function in L^2 and f_n is a given orthogonal sequence in L^2. How can we compute the coefficients c_n? The demonstration here is reminiscent of that given in Section 3.1 where we showed how to compute the coefficients in a sine series. Orthogonality is crucial. We will assume in this argument that the series in (3.12) can be integrated term by term (it can be proved). If we multiply both sides of (3.12) by f_m and integrate over the interval $[a, b]$, we get, using the inner product notation,

$$
\begin{aligned}
(f, f_m) &= \int_a^b \sum_{n=1}^{\infty} c_n f_n(x) f_m(x)\,dx \\
&= \sum_{n=1}^{\infty} c_n (f_n, f_m) \\
&= c_m (f_m, f_m) = c_m \|f_m\|^2.
\end{aligned}
$$

Note that the infinite sum collapsed to a single term because of the orthogonality of the f_n. So the c_n (change the subscript m to n) are given by

$$c_n = \frac{1}{\|f_n\|^2}(f, f_n), \quad n = 1, 2, \ldots . \tag{3.13}$$

Therefore, if (3.12) holds in the mean-square sense, then the coefficients c_n are given by formula (3.13). The c_n are called the *Fourier coefficients*, and when the coefficients are computed in this way, the orthogonal series (3.12) is called the *generalized Fourier series* for f.

If the series expansion (3.12) is valid in the mean-square sense and the f_n are *orthonormal*, then the Fourier coefficients (3.13) are given simply by

$$c_n = (f, f_n), \quad (f_n \text{ orthonormal}). \tag{3.14}$$

There is another useful property of Fourier coefficients. If $f(x)$ is a given function and $f_1, \ldots, f_N$ is an orthonormal set of functions, we can ask how to best approximate f by a linear combination of the f_n. If a_n is any other sequence of coefficients, then

$$\|f - \sum_{n=1}^{N} c_n f_n\|^2 \leq \|f - \sum_{n=1}^{N} a_n f_n\|^2. \tag{3.15}$$

That is, the mean-square error is minimized for the approximation with the Fourier coefficients c_n given by (3.14).

The demonstration of (3.15) is straightforward using the definition of norm and properties of the inner product. We have

$$\begin{aligned}
\|f - \sum_{n=1}^{N} a_n f_n\|^2 &= (f - \sum_{n=1}^{N} a_n f_n, f - \sum_{n=1}^{N} a_n f_n) \\
&= (f, f) - 2\sum_{n=1}^{N} a_n (f, f_n) + \sum_{n=1}^{N} a_n^2 \\
&= (f, f) - 2\sum_{n=1}^{N} a_n c_n + \sum_{n=1}^{N} a_n^2 + \sum_{n=1}^{N} c_n^2 - \sum_{n=1}^{N} c_n^2 \\
&= (f, f) - \sum_{n=1}^{N} c_n^2 + \sum_{n=1}^{N} (a_n - c_n)^2 \\
&= \|f - \sum_{n=1}^{N} c_n f_n\|^2 + \sum_{n=1}^{N} (a_n - c_n)^2.
\end{aligned}$$

Since the second term on the right is nonnegative, (3.15) is valid. If the f_n are not orthonormal, and just orthogonal, then the same result is true with the c_n given by (3.13).

In the last line of the proof we used the fact that

$$\|f - \sum_{n=1}^{N} c_n f_n\|^2 = \|f\|^2 - \sum_{n=1}^{N} c_n^2,$$

which the reader should verify. But this equality leads to another interesting inequality. Since the left side is nonnegative, we have

$$\sum_{n=1}^{N} c_n^2 \leq \|f\|^2.$$

This inequality is valid for each N, and so

$$\sum_{n=1}^{\infty} c_n^2 \leq \|f\|^2,$$

which is *Bessel's inequality*. This shows that the series of squared Fourier coefficients converges, and thus $c_n \to 0$ as $n \to \infty$. So the Fourier coefficients get smaller and smaller.

EXAMPLE

Note that the set of functions

$$\sin nx$$

forms an orthogonal set of functions on $[0, \pi]$. Find the best approximation of the function $f(x) = 8x(\pi - x)e^{-x^2}$ of the form

$$c_1 \sin x + c_2 \sin 2x + c_3 \sin 3x + c_4 \sin 4x + c_5 \sin 5x.$$

We know that the c_n are the Fourier coefficients given by (3.13):

$$c_n = \frac{(f, \sin nx)}{\|\sin nx\|^2} = \frac{2}{\pi}\int_0^{\pi} f(x)\sin nx dx.$$

Using a calculator (TI-86) we obtain

$$c_1 = 3.708,\quad c_2 = 4.040,\quad c_3 = 2.230,\quad c_4 = 0.798,\quad c_5 = 0.295.$$

The reader should plot $f(x)$ and the Fourier approximation on the interval $[0, \pi]$ to see how much the two differ. □

If the f_n form an orthonormal set and we form the generalized Fourier series (3.12) with c_n the Fourier coefficients given by (3.14), does the series automatically converge to f in one sense or another? The answer to the question is yes, in the mean-square sense, provided that the orthonormal sequence f_n is complete. An orthonormal system f_n is said to be *complete* if $(f, f_n) = 0$ for all n implies that $f \equiv 0$. Thus, an orthonormal sequence f_n is complete if the only function having all its Fourier coefficients vanish is the zero function. Sometimes it is difficult to show completeness, and we shall usually just state whether a given orthonormal sequence in complete. It is shown in advanced treatments that completeness is equivalent to the property that *any* function $f \in L^2$ can be expanded in a generalized Fourier series (3.12), where the c_n are the Fourier coefficients, and that the series converges to f in the mean-square sense. Furthermore, it can be proved that completeness is equivalent to strict equality holding in the Bessel inequality, i.e., f_n is complete if and only if for each $f \in L^2$,

$$\sum_{n=1}^{\infty} c_n^2 = \|f\|^2.$$

This equation is called *Parseval's equality*. Pointwise convergence results, or even stronger uniform convergence results, are more difficult to obtain.

We can extend the modal interpretation of the generalized Fourier series still further. We can think of f as a signal and the orthonormal set f_n as fundamental modes. The Fourier coefficient c_n determines the contribution of the nth mode, and the generalized Fourier series is the decomposition of the signal into fundamental modes. The sequence of squared coefficients, $c_1^2, c_2^2, c_3^2, \ldots$, is called the *energy spectrum*, and c_n^2 is called the energy of the nth mode; by Parseval's equality, the *total energy* in the signal is $\|f\|^2$.

Exercises

1. Using a table of integrals or a computer algebra package, verify that the set of functions $1, \cos x, \cos 2x, \ldots$ form an orthogonal set on the interval $[0, \pi]$.

Next verify that the set of functions $\cos(n\pi x/l)$, $n = 0, 1, 2 \ldots$, form an orthogonal set on the interval $[0, l]$. If

$$f(x) = \sum_{n=0}^{\infty} c_n \cos(n\pi x/l)$$

in the mean-square on $[0, l]$, what are the formulae for the c_n?

2. (Gram–Schmidt orthogonalization) We know from elementary linear algebra that any set of linearly independent vectors may be turned into a set of orthogonal vectors by the Gram–Schmidt orthogonalization process. The same process works for functions in $L^2[a, b]$. Let $f_1, f_2, f_3, \ldots$ be an independent set of functions in L^2. Define the sequence g_n by $g_1 = f_1$, $g_2 = f_2 - \frac{(f_2,g_1)}{\|g_1\|^2} g_1$, $g_3 = f_3 - \frac{(f_3,g_2)}{\|g_2\|^2} g_2 - \frac{(f_3,g_1)}{\|g_1\|^2} g_1, \ldots$. Show that g_n is an orthogonal sequence.

3. The functions $1, x, x^2, x^3$ are independent functions on the interval $[-1, 1]$. Use Exercise 2 to generate a sequence of four orthogonal polynomials on $[-1, 1]$. Denote the polynomials by $P_0(x), \ldots, P_3(x)$ (they are called *Legendre polynomials*). Find an approximation of e^x on $[-1, 1]$ of the form

$$e^x \approx c_0P_0(x) + c_1P_1(x) + c_2P_2(x) + c_3P_3(x)$$

that is best in the mean-square sense, and graph e^x and the approximation on a set of coordinate axes. What is the pointwise error? What is the maximum pointwise error over $[-1, 1]$. What is the mean-square error?

4. For which powers r is the function $f(x) = x^r$ in $L^2[0, 1]$? In $L^2[0, \infty)$?

5. Let f be defined and integrable on $[0, l]$. The orthogonal expansion

$$\sum_{n=1}^{\infty} b_n \sin\frac{n\pi x}{l}, \quad b_n = \frac{2}{l}\int_0^l f(x)\sin\frac{n\pi x}{l}\,dx,$$

is called the *Fourier sine series* for f on $[0, l]$. Find the Fourier sine series for $f(x) = \cos x$ on $[0, \pi/2]$. What is the Fourier sine series of $f(x) = \sin x$ on $[0, \pi]$?

6. (Hermite functions) Consider the differential equation

$$-y'' + x^2y = Ey, \quad x \in \mathbf{R}, \quad E = \text{constant}, \tag{3.16}$$

and the functions

$$H_n(x) = (-1)^n e^{x^2}\frac{d^n}{dx^n}e^{-x^2}, n = 0, 1, 2, \ldots.$$

(a) Find $H_0(x), \ldots, H_4(x)$. Why is $H_n(x)$ always a polynomial? (These are called *Hermite polynomials*.)

(b) Verify that $v_n(x) = H_n(x)e^{-x^2/2}$ is a solution of (3.16) when $E = 2n + 1$. Hint: You will need to show $H_n' = 2nH_{n-1}$.

(c) Show that $\int_{-\infty}^{\infty} v_n v_m dx = 0$, $m \neq n$, and thus the v_n are orthogonal on the interval $(-\infty, \infty)$. Hint: Use the fact that both v_n and v_m satisfy (3.16) for appropriate E.

(d) If a function $f(x)$ can be represented by $f(x) = \sum_{n=0}^{\infty} c_n v_n(x)$, how would you expect to find the c_n? Assume uniform convergence. Take $f(x) = 1/\sqrt{1 + x^4}$ and use your calculator to find $c_0, \ldots, c_4$.

(e) The differential equation (3.16) is a time-independent Schrödinger equation (see Section 1.5). Thus $E = 2n + 1$ can be interpreted as the energy associated with the probability density $v_n(x)^2$, appropriately normalized so that $\int v_n^2 dx = 1$. Sketch the normalized probability densities for $n = 0, 1, 2$.

7. (The Haar wavelets) In this exercise we explore the properties of a complete orthonormal system on $(-\infty, \infty)$ called the Haar wavelets. Let ϕ be a function such that $\phi(x) = 1$ for $x \in [0, 1)$, and $\phi(x) = 0$ otherwise. Let $\psi(x) = \phi(2x) - \phi(2x - 1)$. Then the Haar wavelets are the functions

$$\psi_{mn}(x) = 2^{m/2}\psi(2^m x - n),$$

for $m, n = 0, \pm 1, \pm 2, \ldots$. Sketch a graph of $\psi(x)$, and then sketch a graph of $\psi_{mn}(x)$ for $m, n = 0, \pm 1, \pm 2$. What is the graph of $\psi_{mn}(x)$? If f is square integrable and

$$f(x) = \sum_{m=-\infty}^{\infty} \sum_{n=-\infty}^{\infty} c_{mn}\psi_{mn}(x),$$

find a simple formula for the coefficients c_{mn}.

8. For $f, g \in L^2[a, b]$, prove the *Cauchy-Schwarz inequality*

$$|(f, g)| \leq ||f||\, ||g||.$$

Hint: Proceed as follows. Define $q(t) = (f + tg, f + tg)$ for any real number t. Using the rules of scalar product, expand this expression and obtain a quadratic polynomial in t; because $q(t) \geq 0$ (why?), the quadratic polynomial can have at most one real root. Look at the discriminant of the polynomial.

3.3 Classical Fourier Series

In the last section we introduced the idea of representing a given function $f(x)$ in terms of an infinite series of orthogonal functions $f_n(x)$. Now we want to focus those concepts on a special set of orthogonal functions where the f_n are given by sines and cosines. The resulting series is called a (classical) Fourier series. Such series were introduced by Joseph Fourier in the early 1800s. Classical Fourier series are a special case of the orthogonal series discussed in the last section. We will usually drop the adjective "classical."

Equation (3.11) is an example of a Fourier series on the interval $[-\pi, \pi]$. We will work on an arbitrary symmetric interval $[-l, l]$ about the origin. If f is an integrable function on $[-l, l]$, then its *Fourier series* is

$$\frac{a_0}{2} + \sum_{n=1}^{\infty}(a_n \cos \frac{n\pi x}{l} + b_n \sin \frac{n\pi x}{l}), \tag{3.17}$$

where the *Fourier coefficients* a_n and b_n are given by the formulae

$$a_n = \frac{1}{l}\int_{-l}^{l} f(x) \cos \frac{n\pi x}{l}\, dx, \quad n = 0, 1, 2, \ldots, \tag{3.18}$$

$$b_n = \frac{1}{l}\int_{-l}^{l} f(x)\sin\frac{n\pi x}{l}\,dx, \quad n = 1, 2, \ldots. \tag{3.19}$$

Here, the set of functions

$$\frac{1}{2}, \ \cos\frac{n\pi x}{l}, \ \sin\frac{n\pi x}{l}, \qquad n = 1, 2, \ldots, \tag{3.20}$$

are orthogonal on the interval $[-l, l]$; they are playing the role of the $f_n(x)$ in formula (3.12), and the a_n and b_n are playing the role of the coefficients c_n. It is shown in more advanced texts that the set of functions (3.20) is complete, and therefore the Fourier series (3.17) converges in the mean-square sense to f when $f \in L^2[-l, l]$.

EXAMPLE

Let $f(x) = x$ for $-l \le x \le l$. Then the Fourier coefficients are easily computed to be

$$a_n = \frac{1}{l}\int_{-l}^{l} x\cos\frac{n\pi x}{l}\,dx = 0, \quad n = 0, 1, 2, \ldots,$$

$$b_n = \frac{1}{l}\int_{-l}^{l} x\sin\frac{n\pi x}{l}\,dx = \frac{2l}{n\pi}(-1)^{n+1}, \quad n = 1, 2, \ldots.$$

So the Fourier series for f is

$$\frac{2l}{\pi}\left(\sin\frac{\pi x}{l} - \frac{1}{2}\sin\frac{2\pi x}{l} + \frac{1}{3}\sin\frac{3\pi x}{l} - \cdots\right).$$

We make two observations. First, at $x = \pm l$ the series clearly converges to zero and thus does not converge to $f(x)$ at these two points. So the series does not converge pointwise to $f(x)$ on the interval $[-l, l]$. It does, however, converge pointwise to $f(x) = x$ on $(-l, l)$. Second, the derived series, obtained by differentiating term by term, does not converge at all, much less to the derivative $f'(x) = 1$. So the series cannot be differentiated term by term. We do know that the series converges to f in the mean-square sense. □

For the Fourier series (3.17) we define the nth *harmonic* to be

$$a_n \cos\frac{n\pi x}{l} + b_n \sin\frac{n\pi x}{l}.$$

Notice that each harmonic in the series has a higher frequency (and thus more oscillations) than the preceding harmonic. The *frequency spectrum* of the series is the sequence of numbers γ_n defined by

$$\gamma_0 = \frac{|a_0|}{\sqrt{2}}, \quad \gamma_n = \sqrt{a_n^2 + b_n^2} \quad (n \ge 1).$$

The frequency spectrum is a measure of the contribution of the various harmonics in the decomposition of f. The numbers γ_n^2, $n \ge 0$, form the *energy spectrum*. The reader is invited to show that Parseval's formula takes the form

$$\frac{a_0^2}{2} + \sum_{n=1}^{\infty}(a_n^2 + b_n^2) = \frac{1}{l}\,||f||^2.$$

The frequency spectrum for $f(x) = x$ in the last example is $\gamma_0 = 0, \gamma_n = 2l/(n\pi), \quad n \geq 1$. It is often graphed, as in Figure 3.1, to show visually the contribution of each harmonic.

A Fourier series simplifies considerably if f is an even or an odd function. First observe that $\sin \frac{n\pi x}{l}$ is an odd function and $\cos \frac{n\pi x}{l}$ is an even function; moreover, an even function times an odd function is odd. Therefore, if f is an even function, then the product $f(x)\sin \frac{n\pi x}{l}$, which is the integrand in the expression for b_n in (3.19), is an odd function; so all of the coefficients b_n are zero because an odd function integrated over a symmetric interval about the origin is zero. Hence, if f is an even function, then its Fourier series reduces to a series of the form

$$\frac{a_0}{2} + \sum_{n=1}^{\infty} a_n \cos \frac{n\pi x}{l}.$$

Similarly, if f is an odd function, then the coefficients a_n in (3.18) vanish, and the Fourier series reduces to

$$\sum_{n=1}^{\infty} b_n \sin \frac{n\pi x}{l}.$$

This fact is illustrated in the last example, where we found the Fourier series for $f(x) = x$, an odd function.

We commented earlier that Fourier series are a valuable tool in signal processing and data storage. Let us explore this idea in more detail. For this discussion it is helpful to think of the variable x as representing time. We observe that the Fourier series (3.17), although computed only on the

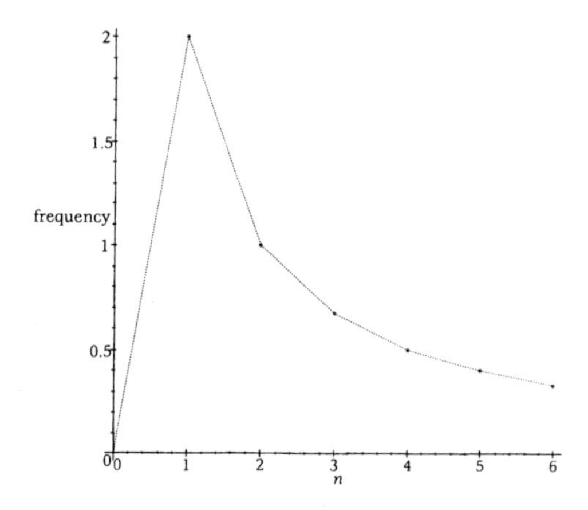

Figure 3.1. Frequency spectrum for the function $f(x) = x$ on $[-\pi, \pi]$.

interval $[-l, l]$, is a $2l$-periodic function* because the sines and cosines are $2l$-periodic. Thus, it repeats its values in every interval of length $2l$. Therefore, if f is defined on all of $\mathbf{R}$ and is $2l$-periodic, we can represent it by its Fourier series on all of $\mathbf{R}$. Hence, if f is some given signal or data set $2l$-periodic in time (say a signal from an electrical circuit, an electrocardiogram, or an extraterrestial), then we could digitize the signal and save it by storing the Fourier coefficients or some finite subset of them; knowledge of those coefficients would allow us reproduce the signal via (3.17)–(3.19).

EXAMPLE

Consider the 2-periodic signal shown in Figure 3.2, which is called a triangular wave. Analytically it is given by $f(x) = x + 1$ if $-1 < x \leq 0$; $f(x) = 1 - x$ if $0 < x \leq 1$; and otherwise 2-periodic. We compute its Fourier series. Here, f is an even function, and so $b_n = 0$ for all n. The coefficients a_n are given by

$$\begin{aligned} a_n &= \int_{-1}^{1} f(x) \cos n\pi x dx \\ &= \int_{-1}^{0} (x+1) \cos n\pi x dx + \int_{0}^{1} (1-x) \cos n\pi x dx. \end{aligned}$$

When $n = 0$, we easily get $a_0 = 1$. For $n \geq 1$ we can calculate a_n by hand (using integration by parts), use a calculator, or use a computer algebra package. The next five coefficients are

$$a_1 = 0.405, \ a_2 = 0, \ a_3 = 0.045, \ a_4 = 0, \ a_5 = 0.016.$$

Therefore, a four-term Fourier approximation to the triangular wave is

$$f(x) \approx 0.5 + 0.405 \cos \pi x + 0.045 \cos 3\pi x + 0.016 \cos 5\pi x.$$

Figure 3.2 shows a plot of f and its approximation. From the last section we know that this approximation is the best, in the mean-square sense. □

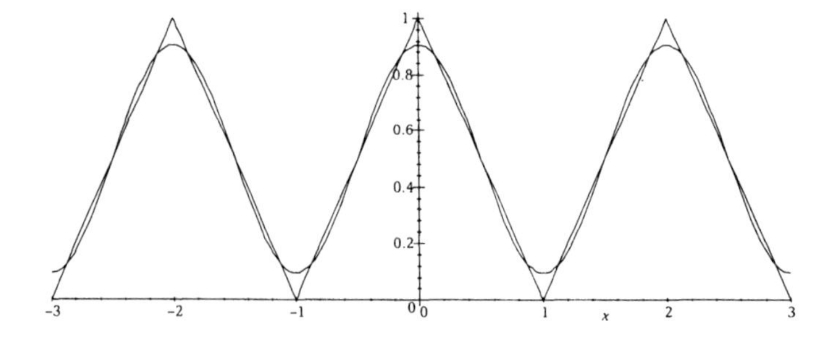

Figure 3.2. A triangular wave.

*Recall that a function f is P-periodic, or periodic of period P, if it repeats itself in every interval of length P; that is, $f(x + P) = f(x)$ for all x.

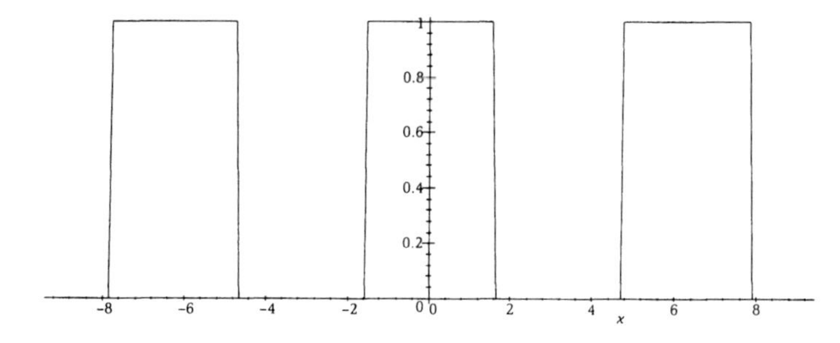

Figure 3.3. Graph of the 2π-periodic square wave.

Finally, we make some brief comments about convergence of Fourier series. We have already noted that if f is square integrable, then mean-square convergence is automatic. To obtain pointwise convergence results some additional assumptions must be made regarding the smoothness of f. Our assumption is that the graph of f is made up of finitely many, not necessarily continuous, smooth pieces; this will cover most of the interesting functions encountered in science and engineering. A function f is *piecewise continuous* on $[a, b]$ if it is continuous except possibly at a finite number of of points in $[a, b]$ where it has simple jump discontinuities; f has a simple jump discontinuity at $x = c$ if both one-sided limits of $f(x)$ exist and are finite at c. The function may or may not be defined at a jump discontinuity. We say f is *piecewise smooth* on $[a, b]$ if both f and f' are piecewise continuous on $[a, b]$, and we say f is piecewise smooth on $(-\infty, \infty)$ if it is piecewise smooth on each bounded subinterval $[a, b]$ of the real line. Then the basic pointwise convergence theorem is the following:

Theorem
If f is piecewise smooth on $[-l, l]$ and otherwise $2l$-periodic, then its Fourier series (3.17) converges pointwise for all $x \in \mathbf{R}$ to the value $f(x)$ if f is continuous at x, and to the average value of its left and right limits at x, namely $\frac{1}{2}(f(x-) + f(x+))$, if f is discontinuous at x.

To get stronger convergence results, like uniform convergence, additional smoothness conditions on f are required. Note that continuity of f is not enough to guarantee pointwise convergence; incredible as it may seem, there are continuous functions whose Fourier series diverge at every point!

Exercises

1. Find the Fourier series for the 2π-periodic square wave shown in Figure 3.3. Sketch a two-term, a four-term, and a six-term approximation.

2. Find the Fourier series for $f(x) = x^2$ on $[-\pi, \pi]$. Sketch a graph of the function defined on all of **R** to which the Fourier series converges. Is the convergence pointwise? Use the Fourier series to show

$$\frac{\pi^2}{12} = 1 - \frac{1}{4} + \frac{1}{9} - \frac{1}{16} + \cdots.$$

Graph the frequency spectrum.

3. Consider the function (signal) defined by $f(x) = x + 1$ for $-2\pi < x \leq 0$ and $f(x) = x$ for $0 < x \leq 2\pi$, and otherwise 4π periodic. Sketch a graph of the signal and find its Fourier series. Find and graph the frequency spectrum. To what value does the Fourier series converge at $x = 0$? At $x = 2\pi$? At $x = \pi$? Graph the sum of the first four harmonics and observe how well it approximates f.

4. Show that the Fourier series for $f(x) = \cos ax$ on $[-\pi, \pi]$, where a is not an integer, is

$$\cos ax = \frac{2a \sin a\pi}{\pi} \left(\frac{1}{2a^2} + \sum_{n=1}^{\infty} (-1)^n \frac{\cos nx}{a^2 - n^2} \right).$$

Show

$$\csc a\pi = \frac{1}{a\pi} + \frac{2a}{\pi} \sum_{n=1}^{\infty} (-1)^n \frac{1}{n^2 - a^2}.$$

5. Let $f(x) = -\frac{1}{2}$ on $-\pi < x \leq 0$ and $f(x) = \frac{1}{2}$ on $0 \leq x \leq \pi$. Show that the Fourier series for f is

$$\sum_{n=1}^{\infty} \frac{2}{(2n-1)\pi} \sin(2n-1)x.$$

If $s_N(x)$ denotes the sum of the first N terms, sketch a graph of $s_1(x)$, $s_3(x)$, $s_7(x)$, and $s_{10}(x)$ and compare with $f(x)$. Observe that the approximations overshoot $f(x)$ in a neighborhood of $x = 0$, and the overshoot is not improved regardless of how many terms are taken in the approximation. This overshoot behavior of Fourier series near a discontinuity is called the *Gibbs phenomenon*.

3.4 Sturm–Liouville Problems

Let us revisit the basic idea of Fourier that we discussed in Section 3.1. We were able to represent the solution of the initial boundary value problem

$$u_t = u_{xx}, \quad 0 < x < \pi, \ t > 0, \tag{3.21}$$

$$u(0, t) = u(\pi, t) = 0, \quad t > 0, \tag{3.22}$$

$$u(x, 0) = f(x), \quad 0 < x < \pi, \tag{3.23}$$

as

$$u(x,t) = \sum_{n=1}^{\infty} c_n e^{-n^2 t} \sin nx \tag{3.24}$$

for suitably chosen constants c_n. We note that this solution is an expansion in terms of orthogonal functions $\sin nx$. We recall from Section 3.1 that orthogonality was essential in computing the c_n from the initial condition. How do we know that this particular set of orthogonal functions is the appropriate choice for this problem?

In answering this question we now describe one of the fundamental methods, called the *Fourier method* or *separation of variables*, to solve PDEs on bounded domains. We will observe that every such PDE leads in a natural way to a boundary value problem for an ODE whose solutions are the appropriate orthogonal functions for the problem; this will mean that the solution to the PDE can be represented in terms of those orthogonal functions.

We restrict our attention to (3.21)–(3.22). The separation of variables method consists in looking for solutions of the form of products, i.e.,

$$u(x,t) = y(x)g(t). \tag{3.25}$$

Substituting this into the PDE (3.21) and boundary conditions (3.22), we obtain

$$y(x)g'(t) = y''(x)g(t), \quad y(0)g(t) = 0, \quad y(\pi)g(t) = 0,$$

which we can write as

$$\frac{g'(t)}{g(t)} = \frac{y''(x)}{y(x)} = -\lambda, \quad y(0) = y(\pi) = 0.$$

Here, $-\lambda$ is some unknown constant (we may set g'/g and y''/y equal to a common constant because the only way a function of t can equal a function of x for all t and x is if both are equal to a constant); the minus sign on λ is for convenience. Therefore, we obtain an ordinary differential equation for g, namely,

$$g'(t) = -\lambda g(t),$$

and we obtain a boundary value problem for y, namely,

$$-y''(x) = \lambda y(x), \quad 0 < x < \pi, \tag{3.26}$$

$$y(0) = 0, \quad y(\pi) = 0. \tag{3.27}$$

So, the PDE problem separated into two ODE problems, for $g(t)$ and $y(x)$. We can easily solve the g-problem; it has a solution

$$g(t) = e^{-\lambda t}.$$

As it turns out (our goal is to discuss this in detail), the boundary value problem (3.26)–(3.27) for y will have a solution only when the constant λ

takes on the special values $\lambda = \lambda_n = n^2$, $n = 1, 2, \ldots$. The corresponding solutions are $y = y_n(x) = \sin nx$, $n = 1, 2, \ldots$. Then we will have $g = g_n(t) = e^{-n^2 t}$, $n = 1, 2, \ldots$. Thus we will have constructed infinitely many solutions

$$u = u_n(x, t) = g_n(t) y_n(x) = e^{-n^2 t} \sin nx$$

that satisfy the PDE (3.21) and the boundary conditions (3.22). Now we can clearly see where the terms in the solution (3.24) arise. The orthogonal functions $\sin nx$ come from the solution of the ODE boundary value problem (3.26)–(3.27) for y.

What we have just briefly described is the separation of variables method, and the procedure can be imitated on a large number of PDE models. Each will give rise to a boundary value problem for an ODE that has infinitely many solutions that form an orthogonal system. In this section we focus attention on the ODE boundary value problem, saving a full discussion of the separation of variables method to Chapter 4.

When we separate variables, i.e., assume (3.25), for an equation of the form

$$u_t = (p(x)u_x)_x - q(x)u, \quad a < x < b, \ t > 0$$

(here, a and b are finite), we obtain an ODE for $y = y(x)$ of the form

$$-(p(x)y')' + q(x)y = \lambda y, \quad a < x < b, \tag{3.28}$$

where λ is some constant (see Exercise 1). This ODE is called a *Sturm–Liouville* differential equation. The boundary conditions on the PDE will usually lead to boundary conditions on $y(x)$ of the form

$$\alpha_1 y(a) + \alpha_2 y'(a) = 0, \tag{3.29}$$

$$\beta_1 y(b) + \beta_2 y'(b) = 0. \tag{3.30}$$

Here the constants α_1 and α_2 are not both zero, and the constants β_1 and β_2 are not both zero (i.e., the boundary condition at an endpoint does not collapse). Two special cases of the boundary conditions (3.29)–(3.30) are

$$y(a) = 0, \quad y(b) = 0 \quad \text{(Dirichlet conditions)} \tag{3.31}$$

and

$$y'(a) = 0, \quad y'(b) = 0 \quad \text{(Neumann conditions).} \tag{3.32}$$

If p, p', and q are continuous functions on the interval $[a, b]$ and p is never zero in $[a, b]$, then the ODE boundary value problem (3.28)–(3.30) is called a *regular* Sturm–Liouville problem (SLP). Otherwise it is called *singular*. Singular problems usually arise when the interval $[a, b]$ is infinite or when $p(x_0) = 0$ for some $x_0 \in [a, b]$. SLPs are named after J.C.F. Sturm and J. Liouville, who studied such problems in the mid 1830s.

A regular SLP will not have a solution for every value of the constant λ. (Clearly, the zero function $y(x) \equiv 0$ is always a trivial solution, but we have no interest in it). A value of λ for which there is a nontrivial solution of (3.28)–(3.30) is called an *eigenvalue*, and the corresponding solution is called an *eigenfunction*; observe that any constant multiple of an eigenfunction gives another (not independent) eigenfunction. The interesting fact about regular SLPs is that they have an infinite number of eigenvalues, and the corresponding eigenfunctions form a complete, orthogonal set, which makes orthogonal expansions possible. This is a key idea in applied mathematics.

Before introducing some general facts for SLPs, we study some examples.

EXAMPLE

The SLP

$$-y''(x) = \lambda y(x), \quad 0 < x < \pi, \tag{3.33}$$

$$y(0) = 0, \quad y(\pi) = 0, \tag{3.34}$$

has eigenvalues $\lambda = \lambda_n = n^2$ and corresponding eigenfunctions $y = y_n(x) = \sin nx$, $n = 1, 2, \ldots$. Thus (3.33)–(3.34) has solution $y = \sin x$ when $\lambda = 1$, $y = \sin 2x$ when $\lambda = 4$, and so on. One way to find the eigenvalues and eigenfunctions is to go through a case argument by separately considering $\lambda = 0, \lambda > 0$, and $\lambda < 0$. We will prove later that λ cannot be a complex number. We examine different cases because the solution of (3.33) has a different form depending on the sign of λ. If $\lambda = 0$, then the ODE (3.33) has the form $y'' = 0$, whose general solution is the linear function $y(x) = Ax + B$. But $y(0) = 0$ implies $B = 0$, and $y(\pi) = 0$ implies $A = 0$. Thus we get only the trivial solution in this case and so $\lambda = 0$ is *not* an eigenvalue. If $\lambda < 0$, say for definiteness $\lambda = -k^2$, then the ODE (3.33) has the form $y'' - k^2 y = 0$; this has general solution consisting of exponentials,

$$y(x) = Ae^{kx} + Be^{-kx}.$$

The boundary conditions force

$$y(0) = A + B = 0,$$

$$y(\pi) = Ae^{k\pi} + Be^{-k\pi} = 0.$$

It is easy to see that these two equations for A and B force $A = B = 0$, and therefore we obtain only the trivial solution in this case. Thus, there are no negative eigenvalues. Finally, let us consider the case $\lambda > 0$, or $\lambda = k^2$. Then the ODE (3.33) takes the form

$$y'' + k^2 y = 0,$$

which has solutions built of sines and cosines. That is,

$$y(x) = A\cos kx + B\sin kx.$$

First, $y(0) = 0$ forces $A = 0$. Thus $y(x) = B \sin kx$. The right-hand boundary condition yields

$$y(\pi) = B \sin k\pi = 0.$$

But now we are not required to take $B = 0$; rather, we can select k to make this equation hold. Clearly, since the sine function vanishes at multiples of π, we can choose $k = n$, a nonzero integer (we have already considered the case when $k = 0$), or

$$\lambda = \lambda_n = n^2, \quad n = 1, 2, \ldots.$$

The corresponding solutions, or eigenfunctions, are

$$y = y_n(x) = \sin nx, \quad n = 1, 2, \ldots.$$

Here we have arbitrarily selected the constant $B = 1$ for each eigenfunction; we can always multiply eigenfunctions by a constant to get another (not independent) eigenfunction. So, frequently one sees eigenfunctions written as, for example, $y_n(x) = B_n \sin nx$, where a different constant is chosen for each eigenfunction. Sometimes the constants B_n are chosen to normalize the eigenfunction, i.e., force $||y_n|| = 1$. As we shall see later, it does not matter how we choose the B_n at this point, because we shall end up taking an arbitrary linear combination of the eigenfunctions. □

We can always carry out a case argument as in the preceding example to determine the eigenvalues and eigenfunctions. This calculation can be tedious, and so it will be advantageous to have some general results that allow us to reject certain cases immediately, without detailed calculation. We discuss such matters in the sequel. For the present we state, without proof, the central result regarding SLPs.

Theorem

The regular Sturm–Liouville problem (3.28)–(3.30) has infinitely many eigenvalues λ_n, $n = 1,2,\ldots$. The eigenvalues are real, and $\lim_{n\to\infty} |\lambda_n| = +\infty$. The eigenfunctions corresponding to distinct eigenvalues are orthogonal, and the set of all eigenfunctions $y_n(x)$ is complete in the sense that every square-integrable function f on $[a,b]$ can be expanded as

$$f(x) = \sum_{n=1}^{\infty} c_n y_n(x)$$

in the mean-square sense.

It is not particularly easy to prove the existence of the eigenvalues (see, for example, Birkhoff and Rota [1965]). But it is straightforward to demonstrate the orthogonality. To this end, let λ_1 and λ_2 be distinct eigenvalues with corresponding eigenfunctions $y_1(x)$ and $y_2(x)$. This means

$$-(py_1')' + qy_1 = \lambda_1 y_1,$$
$$-(py_2')' + qy_2 = \lambda_2 y_2.$$

Multiplying the first equation by y_2, the second by y_1, subtracting, and then integrating the result over the interval $[a, b]$, gives

$$(\lambda_1 - \lambda_2)\int_a^b y_1 y_2 dx = \int_a^b [-y_2(py_1')' + y_1(py_2')']dx.$$

The integral on the right can be calculated by noticing that the integrand is an exact derivative; i.e.,

$$\frac{d}{dx}[p(y_1 y_2' - y_2 y_1')] = -y_2(py_1')' + y_1(py_2')' \tag{3.35}$$

(see Exercise 5). Then the fundamental theorem of calculus can be applied, and we obtain

$$(\lambda_1 - \lambda_2)\int_a^b y_1 y_2 dx = [p(y_1 y_2' - y_2 y_1'] \,|_a^b \,.$$

For the simple Dirichlet or Neumann boundary conditions (3.31) and (3.32), respectively, it is clear that the right side is zero, since both eigenfunctions satisfy the boundary conditions. For the general boundary conditions (3.29)–(3.30) the right side is zero as well, and we leave this verification to the reader. Thus it follows that

$$\int_a^b y_1 y_2 dx = 0,$$

which means that y_1 and y_2 are orthogonal functions.

We stated earlier that eigenvalues of the regular SLP cannot be complex. We will demonstrate this fact for the special SLP (a Schrödinger-type equation) with Dirichlet boundary conditions

$$-y'' + q(x)y = \lambda y, \quad a < x < b, \tag{3.36}$$
$$y(a) = y(b) = 0. \tag{3.37}$$

Let λ be an eigenvalue and $y(x)$ a corresponding eigenfunction. Let an *overbar* denote complex conjugation. Then, taking the complex conjugate of the ODE (3.36) gives

$$-\overline{y}'' + q(x)\overline{y} = \overline{\lambda}\overline{y}.$$

Here, of course, we are assuming that q is a real-valued function. Then we multiply (3.36) by $\overline{y}$, and we multiply the last equation by y; subtracting the results and integrating then yields

$$-\int_a^b [y''\overline{y} - \overline{y}''y]dx = (\lambda - \overline{\lambda})\int_a^b y\overline{y}dx.$$

Using the identity (3.35), we can show that the integral on the left side is zero:

$$\int_a^b [y''\bar{y} - \bar{y}''y]dx = (y'\bar{y} - \bar{y}'y)\,|_a^b = 0.$$

Here we have used $\bar{y}(a) = \bar{y}(b) = 0$, which comes from taking the complex conjugate of the boundary conditions. Therefore, we have shown that

$$(\lambda - \bar{\lambda})\int_a^b y\bar{y}dx = 0.$$

Because the integral is positive (the integrand $y\bar{y} = |y|^2$ is positive), we must have $\lambda = \bar{\lambda}$. A number that equals its complex conjugate is real, and so λ is real.

We leave as an exercise to show that eigenvalues for the general SLP (3.28)–(3.30) are real.

Another question concerns the sign (positive or negative) of eigenvalues. The following type of *energy argument* is sometimes useful in showing that the eigenvalues are of one sign. We work with (3.36)–(3.37). If we multiply the ODE (3.36) by y and integrate, we get

$$-\int_a^b yy''dx + \int_a^b qy^2dx = \lambda\int_a^b y^2dx.$$

The first integral can be integrated by parts, and we obtain

$$-yy'\,|_a^b + \int_a^b y'^2dx + \int_a^b qy^2dx = \lambda\int_a^b y^2dx.$$

Under Dirichlet boundary conditions (3.37) we see that the boundary terms in the last equation are zero. Hence

$$\int_a^b y'^2dx + \int_a^b qy^2dx = \lambda\int_a^b y^2dx.$$

If q is positive, then the second integral on the left side is positive. Thus λ must be positive.

Exercises

1. Show that substitution of $u(x,t) = g(t)y(x)$ into the PDE

$$u_t = (p(x)u_x)_x - q(x)u, \quad a < x < b,\ t > 0,$$

leads to the pair of differential equations

$$g' = -\lambda g, \quad -(p(x)y')' + q(x)y = \lambda y,$$

where λ is some constant.

2. Show that the SLP

$$-y''(x) = \lambda y(x), \quad 0 < x < l,$$
$$y(0) = 0, \quad y(l) = 0,$$

has eigenvalues $\lambda_n = n^2\pi^2/l^2$ and corresponding eigenfunctions $y_n(x) = \sin(n\pi x/l)$, $n = 1, 2, \ldots$.

3. Show that the SLP

$$-y''(x) = \lambda y(x), \quad 0 < x < l,$$
$$y'(0) = 0, \quad y(l) = 0,$$

with mixed Dirichlet and Neumann boundary conditions has eigenvalues

$$\lambda_n = \left(\frac{(1+2n)\pi}{2l}\right)^2$$

and corresponding eigenfunctions

$$y_n(x) = \cos\frac{(1+2n)\pi x}{2l}$$

for $n = 0, 1, 2, \ldots$.

4. Find the eigenvalues and eigenfunctions for the following problem with *periodic* boundary conditions:

$$-y''(x) = \lambda y(x), \quad 0 < x < l,$$
$$y(0) = y(l), \quad y'(0) = y'(l).$$

5. Verify equation (3.35).

6. Consider the SLP

$$-y'' = \lambda y, \quad 0 < x < 1; \quad y(0) + y'(0) = 0, \quad y(1) = 0.$$

Is $\lambda = 0$ an eigenvalue? Are there any negative eigenvalues? Show that there are infinitely many positive eigenvalues by finding an equation whose roots are those eigenvalues, and show graphically that there are infinitely many roots.

7. Show that the SLP

$$-y'' = \lambda y, \quad 0 < x < 2; \quad y(0) + 2y'(0) = 0, \quad 3y(2) + 2y'(2) = 0,$$

has exactly one negative eigenvalue. Is zero an eigenvalue? How many positive eigenvalues are there?

8. For the SLP

$$-y'' = \lambda y, \quad 0 < x < l; \quad y(0) - ay'(0) = 0, \quad y(l) + by'(l) = 0,$$

show that $\lambda = 0$ is an eigenvalue if and only if $a + b = -l$.

9. Consider the SLP

$$-(x^2y')' = \lambda y, \quad 1 < x < \pi, \quad y(1) = y(\pi) = 0.$$

Use an energy argument as presented in the text to show that any eigenvalue must be nonnegative. Find the eigenvalues and eigenfunctions.

10. What can you say about the sign of the eigenvalues for the SLP

$$-y'' + xy = -\lambda y, \quad 0 < x < 1, \quad y(0) = y(1) = 0.$$

Use a computer algebra package to find the eigenvalues and eigenfunctions. Hint: Look up Airy's differential equation.

11. Consider the regular SLP

$$-y'' + q(x)y = \lambda y, \quad 0 < x < l,$$
$$y(0) = y(l) = 0,$$

where $q(x) > 0$ on $[0, l]$. Show that if λ and y are an eigenvalue and eigenfunction, respectively, then

$$\lambda = \frac{\int_0^l (y'^2 + qy^2)dx}{||y||^2}.$$

Is $\lambda > 0$? Can $y(x) =$ constant be an eigenfunction?

12. Can there be two independent eigenfunctions corresponding to a single eigenvalue for a regular SLP? Think about this question for the SLP given in Exercise 11. What about the problem given in Exercise 4?

13. Find eigenvalues and eigenfunctions for the singular problem

$$-y'' + x^2 y = \lambda y, \quad x \in \mathbf{R}; \quad y \in L^2(\mathbf{R}).$$

See Exercise 6 in Section 3.2.

4 CHAPTER Partial Differential Equations on Bounded Domains

4.1 Separation of Variables

The standard technique for solving PDEs on bounded domains is called *separation of variables*(or the method of *eigenfunction expansions*, or the *Fourier method*). As we have already noted in Section 3.4, the idea is to assume that the unknown function $u = u(x, t)$ (we assume that the independent variables are x and t) in an initial boundary value problem can be written as a product of a function of x and a function of t, that is, $u(x, t) = y(x)g(t)$. Thus, the variables separate. If the method is to be successful, when this product is substituted into the PDE, the PDE separates into two ODEs, one for $y(x)$ and one for $g(t)$. Substitution of the product into the boundary conditions leads to boundary conditions on $y(x)$. Therefore, we are left with an ODE boundary value problem for $y(x)$ and an ODE for $g(t)$. When we solve for $y(x)$ and $g(t)$, we will have a product solution $u(x, t)$ of the PDE that satisfies the boundary conditions. As it turns out, the boundary value problem for $y(x)$ is a Sturm–Liouville problem and will, in fact, have infinitely many solutions; consequently, we will have infinitely many product solutions $u_1(x, t), u_2(x, t), \ldots$ that satisfy the boundary conditions. One can then superimpose these solutions, or add them up in a special way, to determine a solution of the PDE and boundary conditions that *also* satisfies the initial condition(s). In other words, we form the series

$$c_1u_1(x, t) + c_2u_2(x, t) + \cdots$$

and choose the constants such that the combination satisfies the initial condition(s) as well. So the result of the calculation is an infinite series representation of the solution to the initial boundary value problem.

In this section we go through a detailed presentation of the Fourier method.

We apply the procedure to the initial boundary value problem for the heat equation:

$$u_t = ku_{xx}, \quad 0 < x < l, \ t > 0, \tag{4.1}$$

$$u(0,t) = 0, \quad u(l,t) = 0, \quad t > 0, \tag{4.2}$$

$$u(x,0) = f(x), \quad 0 < x < l. \tag{4.3}$$

For the separation of variables method to work, the boundary conditions must be homogeneous.

Step one is to assume a solution of the form $u(x,t) = y(x)g(t)$ and substitute it into the PDE (4.1) and boundary conditions (4.2). Substituting into the PDE gives

$$y(x)g'(t) = ky''(x)g(t),$$

or, upon dividing by kyg,

$$\frac{g'(t)}{kg(t)} = \frac{y''(x)}{y(x)}.$$

Notice that the variables in the equations have been separated—the left side is a function of t, and the right side is a function of x. Now, when can a function of t be equal to a function of x for all $x \in (0, l)$ and all $t > 0$? Only if the two functions are equal to the same constant, which we call $-\lambda$. That is, we must have

$$\frac{g'(t)}{kg(t)} = -\lambda, \qquad \frac{y''(x)}{y(x)} = -\lambda.$$

Therefore, we have two ODEs, for g and y:

$$g'(t) = -\lambda k g(t), \qquad -y''(x) = \lambda y(x).$$

The constant $-\lambda$, called the separation constant, is not yet known. Placing a minus sign on the constant does not mean that it is negative; we have chosen the minus sign to force the ODE for y to be in the standard Sturm–Liouville form (see Section 3.4); for the same reason, we have included the constant k with the $g(t)$ equation. Next we substitute $u(x,t) = y(x)g(t)$ into the boundary conditions (4.2) to obtain

$$y(0)g(t) = 0, \qquad y(l)g(t) = 0.$$

Excluding the uninteresting possibility that $g(t) = 0$, we get $y(0) = 0$ and $y(l) = 0$. Therefore, we are led to the boundary value problem

$$-y''(x) = \lambda y(x), \quad 0 < x < l, \tag{4.4}$$

$$y(0) = 0, \quad y(l) = 0. \tag{4.5}$$

This is a Sturm–Liouville problem for y, and we now recognize the separation constant as an eigenvalue.

The next step is to solve the ODE for $g(t)$ and the Sturm–Liouville problem for $y(x)$. The g-equation can be solved easily (it is the equation for exponential decay) to get

$$g'(t) = -\lambda k g(t) \Rightarrow g(t) = Ce^{-\lambda kt}.$$

The Sturm–Liouville problem (4.4)–(4.5) has already been solved in Section 3.4, Exercise 2. The eigenvalues are

$$\lambda = \lambda_n = \frac{n^2\pi^2}{l^2}, \quad n = 1, 2, \ldots, \tag{4.6}$$

and the corresponding eigenfunctions are

$$y_n(x) = \sin \frac{n\pi x}{l}, \quad n = 1, 2, \ldots. \tag{4.7}$$

To reiterate, the eigenvalues are those values of λ for which the problem (4.4)–(4.5) has a nontrivial solution; the eigenfunctions are those corresponding to nontrivial solutions.

Now we can put together the above results. We have constructed infinitely many product solutions of the PDE (4.1) having the form

$$u_n(x, t) = c_n e^{-n^2\pi^2 kt/l^2} \sin \frac{n\pi x}{l}, \quad n = 1, 2, \ldots,$$

where $\lambda = \lambda_n$ has been substituted into the formula for $g(t)$ and arbitrary constants c_n depending on n have been placed in front of the expression. These product solutions also clearly satisfy the boundary conditions (4.2), but they do not satisfy the given initial condition (4.3). So the next step is to determine the constants c_n such that the linear combination

$$u(x, t) = \sum_{n=1}^{\infty} c_n e^{-n^2\pi^2 kt/l^2} \sin \frac{n\pi x}{l}. \tag{4.8}$$

satisfies the initial condition (4.3). Thus we require

$$u(x, 0) = f(x) = \sum_{n=1}^{\infty} c_n \sin \frac{n\pi x}{l}.$$

But the right side is just the Fourier sine series of the function $f(x)$ on the interval $(0, l)$. Therefore, the coefficients c_n are the Fourier coefficients given by (see Section 3.3)

$$c_n = \frac{2}{l}\int_0^l f(x) \sin \frac{n\pi x}{l}\, dx, \quad n = 1, 2, \ldots. \tag{4.9}$$

Therefore, we have obtained a solution to (4.1)–(4.3) given by the infinite series (4.8) where the coefficients c_n are given by (4.9). If the function f is complicated, then a simple formula for the numbers c_n cannot usually be found, and we must resort to numerical integration to determine these constants.

Substituting the expression for the c_n into the solution formula (4.8) allows us to write the solution as

$$
\begin{aligned}
u(x,t) &= \sum_{n=1}^{\infty}\left(\frac{2}{l}\int_0^l f(\xi)\sin\frac{n\pi\xi}{l}\,d\xi\right)e^{-n^2\pi^2kt/l^2}\sin\frac{n\pi x}{l} \\
&= \int_0^l\left(\frac{2}{l}\sum_{n=1}^{\infty}e^{-n^2\pi^2kt/l^2}\sin\frac{n\pi\xi}{l}\sin\frac{n\pi x}{l}\right)f(\xi)d\xi \\
&\equiv \int_0^l g(x,\xi,t)f(\xi)d\xi.
\end{aligned}
$$

This formula shows nicely the solution as an integral operator acting on the function f.

Several comments are in order. First, the solution we obtained is, indeed, a solution for $0 < x < l$ and $t > 0$, and it satisfies the boundary conditions, for only mild restrictions on the initial temperature distribution $f(x)$ (e.g., f is a piecewise smooth function on $0 \le x \le l$). More severe conditions are required on f if one wishes to prove that the solution surface we obtained is continuous on $0 \le x \le l, \quad t \ge 0$. Justification and verification can be found in more advanced texts on PDEs. Second, one can show that the solution we obtained is the only solution to the problem. Next, for the separation of variables method to be successful, the PDE and boundary conditions must both be linear and homogeneous. The method would not be expected to work for nonlinear equations. If we are faced with a problem having nonhomogeneous boundary conditions, then that problem must be transformed into one with homogeneous conditions; as we observed earlier, that transformation leads to a source term in the PDE. Later we shall learn how to deal with sources. Finally, once an infinite series representation for the solution is found, the work is not over if we want temperature profiles. Generally, we cannot sum an infinite series, and therefore we often take the first few terms as an approximation. Such approximations will be illustrated in an example below and in the exercises.

The separation of variables procedure described above can be imitated for a large number of problems. Now we illustrate the method for the wave equation, but with less detail. Consider the problem

$$u_{tt} = c^2u_{xx}, \quad 0 < x < l, \ t > 0, \tag{4.10}$$

$$u(0,t) = 0, \quad u(l,t) = 0, \quad t > 0, \tag{4.11}$$

$$u(x,0) = F(x), \quad u_t(x,0) = G(x), \quad 0 < x < l. \tag{4.12}$$

Assuming $u(x,t) = y(x)g(t)$ and substituting into the PDE (4.10) yields

$$y(x)g''(t) = c^2y''(x)g(t),$$

or, upon dividing by c^2yg,

$$\frac{g''(t)}{c^2g(t)} = \frac{y''(x)}{y(x)}.$$

Setting each term equal to a constant $-\lambda$, we obtain the two ODEs

$$g''(t) + c^2\lambda g(t) = 0, \quad -y''(x) = \lambda y(x).$$

Next we substitute $u(x, t) = y(x)g(t)$ into the boundary conditions (4.11) to obtain

$$y(0)g(t) = 0, \quad y(l)g(t) = 0,$$

which gives $y(0) = 0$ and $y(l) = 0$. Therefore, we are led to the boundary value problem

$$-y''(x) = \lambda y(x), \quad 0 < x < l; \quad y(0) = 0, \quad y(l) = 0.$$

This is the Sturm–Liouville problem for y, and it is exactly the same as in the heat flow example above. The eigenvalues and eigenfunctions are (see (4.6)–(4.7))

$$\lambda_n = \frac{n^2\pi^2}{l^2}, \quad y_n(x) = \sin\frac{n\pi x}{l}, \quad n = 1, 2 \ldots . \tag{4.13}$$

Solving the ODE for $g(t)$,

$$g(t) = C\sin\frac{n\pi ct}{l} + D\cos\frac{n\pi ct}{l}.$$

Therefore, we have constructed infinitely many product solutions of the PDE (4.10) having the form

$$u_n(x, t) = \left(c_n \sin\frac{n\pi ct}{l} + d_n\cos\frac{n\pi ct}{l}\right)\sin\frac{n\pi x}{l}, \quad n = 1, 2, \ldots,$$

where we have replaced the arbitrary constants C and D by arbitrary constants c_n and d_n depending on n. These product solutions $u_n(x, t)$ represent modes of vibrations. The temporal part is periodic in time with period $2\pi l/(nc\pi)$; the spatial part is a "standing wave" with frequency $n\pi/l$. These product solutions also satisfy the boundary conditions (4.11) but do not satisfy the given initial conditions (4.12). So we form the linear combination

$$u(x, t) = \sum_{n=1}^{\infty}\left(c_n \sin\frac{n\pi ct}{l} + d_n\cos\frac{n\pi ct}{l}\right)\sin\frac{n\pi x}{l} \tag{4.14}$$

and select the constants c_n and d_n such that the initial conditions hold. Thus we require

$$u(x, 0) = F(x) = \sum_{n=1}^{\infty} d_n \sin\frac{n\pi x}{l}.$$

The right side is the Fourier sine series of the function $F(x)$ on the interval $(0, l)$. Therefore, the coefficients d_n are the Fourier coefficients given by

$$d_n = \frac{2}{l}\int_0^l F(x)\sin\frac{n\pi x}{l}\,dx, \quad n = 1, 2, \ldots. \tag{4.15}$$

To apply the other initial condition $u_t(x, 0) = G(x)$ we need to calculate the time derivative of $u(x, t)$. We obtain

$$u_t(x, t) = \sum_{n=1}^{\infty} \frac{nc\pi}{l}\left(c_n \cos\frac{n\pi ct}{l} - d_n \sin\frac{n\pi ct}{l}\right)\sin\frac{n\pi x}{l}.$$

Thus

$$u_t(x, 0) = G(x) = \sum_{n=1}^{\infty} \frac{nc\pi}{l} c_n \sin\frac{n\pi x}{l}.$$

Again, the right side is the Fourier sine series of G on $(0, l)$, so the coefficients of $\sin\frac{n\pi x}{l}$, which are $\frac{nc\pi}{l}c_n$, are the Fourier coefficients; that is,

$$\frac{nc\pi}{l}c_n = \frac{2}{l}\int_0^l G(x)\sin\frac{n\pi x}{l}\,dx, \quad n = 1, 2, \ldots,$$

or

$$c_n = \frac{2}{nc\pi}\int_0^l G(x)\sin\frac{n\pi x}{l}\,dx, \quad n = 1, 2, \ldots. \tag{4.16}$$

Therefore, the solution of the initial boundary value problem is given by the infinite series (4.14) where the coefficients are given by (4.15) and (4.16).

Computer algebra packages can be of tremendous help in developing the formulas found by the separation of variables method to obtain approximate solution surfaces.

Example

Consider the following initial boundary value problem for the heat equation:

$$u_t = u_{xx}, \quad 0 < x < 1,\ t > 0,$$

$$u(0, t) = u(1, t) = 0, \quad t > 0,$$

$$u(x, 0) = 10x^3(1 - x), \quad 0 < x < 1.$$

The solution is given by the formulae (4.8) and (4.9). In Maple the following sequence of commands defines the function $f(x)$, computes the Fourier coefficients c_n, and then finds the first six terms of an approximate

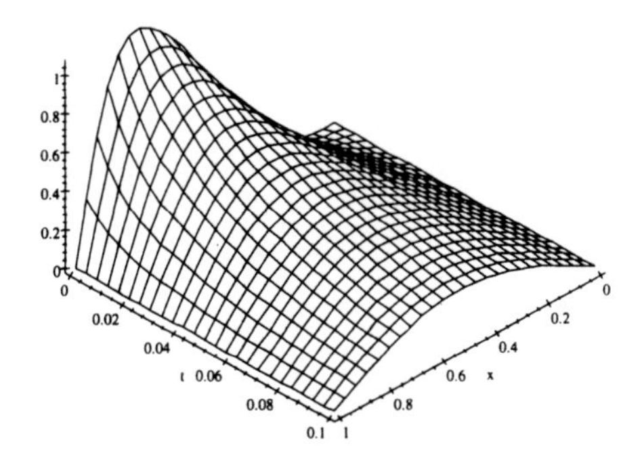

Figure 4.1. The temperature surface.

solution. Here $k = 1$ and $l = 1$.

```
f:=x → 10*x^3*(1-x);
coef:=n → 2*int(f(x)*sin (n*Pi*x),x=0..1);
c:=seq(coef(n),n=1..6);
u:=(x,t) → sum(c[n]*sin(n*Pi*x)
  *exp(-n^2*Pi^2*t),n=1..6);
```

The approximate solution surface can be plotted for $0 \le t \le 0.1$ via the command

```
plot3d(u(x,t),x=0..1,t=0..0.1);
```

The temperature surface is shown in Figure 4.1. The initial temperature profile $f(x)$ and the temperature profile at $t = 0.1$ are shown in Figure 4.2, which was generated by the commands

```
p1:=plot(f(x),x=0..1):
p2:=plot(u(x,0.1),x=0..1):
display([p1,p2]);
```

These graphs show how the bar cools down. □

Exercises

1. In the heat flow problem (4.1)–(4.3) take $k = 1$, $l = \pi$, and $f(x) = 0$ if $0 < x < \pi/2$, $f(x) = 1$ if $\pi/2 < x < \pi$. Find an infinite series representation of the solution. Use the first four terms in the series to obtain an approximate solution, and on the same set of coordinate axes sketch several time snapshots of the approximate temperature distribution in the bar in order to show how the bar cools down. Estimate the error in these approximate distributions.

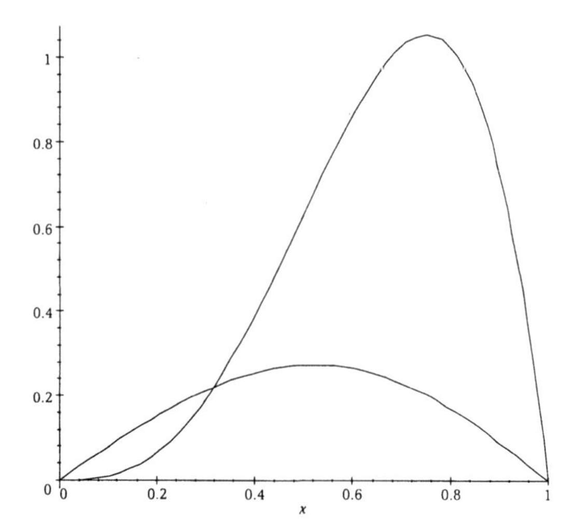

Figure 4.2. Temperature profiles at $t = 0$ and $t = 0.1$.

2. In the vibration problem (4.10)–(4.12) take $c = 1$, $l = \pi$, and $F(x) = x$ if $0 < x < \pi/2$, and $F(x) = \pi - x$ if $\pi/2 < x < \pi$, and take $G(x) \equiv 0$. Find an infinite series representation of the solution. Use the first four terms in the series to obtain an approximate solution, and on the same set of coordinate axes sketch several time snapshots of the wave. Can you estimate the error in these approximate wave forms?

3. Consider the pure boundary value problem for Laplace's equation given by

$$u_{xx} + u_{yy} = 0, \quad 0 < x < l,\ 0 < y < 1,$$
$$u(0, y) = 0, \quad u(l, y) = 0, \quad 0 < y < 1,$$
$$u(x, 0) = F(x), \quad u(x, 1) = G(x), \quad 0 < x < l.$$

Use the separation of variables method to find an infinite-series representation of the solution to this problem. Here, take $u(x, y) = \phi(x)\psi(y)$ and identify a Sturm–Liouville problem for $\phi(x)$; proceed as in the text.

4. The initial boundary value problem for the damped wave equation,

$$u_{tt} + ku_t = c^2 u_{xx}, \quad 0 < x < 1,\ t > 0,$$
$$u(0, t) = 0, \quad u(1, t) = 0, \quad t > 0,$$
$$u(x, 0) = f(x), \quad u_t(x, 0) = 0, \quad 0 < x < 1,$$

governs the displacement of a string immersed in a fluid. The string has unit length and is fixed at its ends; its initial displacement is f, and it has no initial velocity. The constant k is the damping constant. Use separation of variables to find the solution in the case $k < 2\pi c$.

4.2 Flux and Radiation Conditions

In the last section we solved the initial boundary value problem for the heat equation:

$$u_t = ku_{xx}, \quad 0 < x < l, \ t > 0,$$

$$u(0,t) = u(l,t) = 0, \quad t > 0,$$

$$u(x,0) = f(x), \quad 0 < x < l.$$

The boundary conditions in this problem, where we have fixed the temperature to be zero at the ends of the bar, are examples of *Dirichlet*, or fixed endpoint, conditions. Now we wish to consider two other types of boundary conditions; the first is a *Neumann*, or *flux*, condition, and the second is a *Robin*, or *radiation*, condition. An insulation-type flux condition at, say, $x = 0$ is a condition on the derivative of the form $u_x(0,t) = 0, \quad t > 0$; because the flux is proportional to the temperature gradient (Fourier's heat flow law states flux $= -Ku_x$, where K is the conductivity, see Section 1.4), the insulation condition requires that no heat flow across the boundary $x = 0$. A radiation condition, on the other hand, is a specification of how heat radiates from the end of the bar, say at $x = 0$, into its environment, or how the end absorbs heat from its environment. Linear, homogeneous radiation conditions take the form $-Ku_x(0,t) + bu(0,t) = 0, \quad t > 0$, where b is a constant. If $b > 0$, then the heat flux is negative, which means that heat is flowing from the bar into its surroundings (radiation); if $b < 0$, then the flux is positive, and heat is flowing into the bar (absorption). A typical problem in heat conduction may have a combination of Dirichlet, insulation, and radiation boundary conditions. In other physical contexts, e.g., contaminant transport, biological diffusion, these three types of boundary conditions also play an important role.

The method of eigenfunction expansions applies to problems with these different types of homogeneous boundary conditions. The technique is exactly the same as in the last section; the only change will be the form of the boundary conditions in the Sturm–Liouville problem that we obtain upon separating variables.

EXAMPLE

Consider the diffusion problem

$$u_t = ku_{xx}, \quad 0 < x < l, \ t > 0, \tag{4.17}$$

$$u_x(0,t) = 0, \quad u(l,t) = 0, \quad t > 0, \tag{4.18}$$

$$u(x,0) = f(x), \quad 0 < x < l, \tag{4.19}$$

where the left end is insulated and the right end is fixed. We assume that $u(x,t) = y(x)g(t)$ and substitute this into the PDE to obtain, exactly as in

Section 4.1,

$$\frac{g'(t)}{kg(t)} = \frac{y''(x)}{y(x)} = -\lambda,$$

where $-\lambda$ is the separation constant. This gives the two ODEs

$$g'(t) = -\lambda k g(t), \quad -y''(x) = \lambda y(x).$$

Now substituting the product $u = yg$ into the boundary conditions gives

$$y'(0)g(t) = 0, \quad y(l)g(t) = 0.$$

Since we do not want $g(t)$ to be zero, we are forced to take $y'(0) = 0$ and $y(l) = 0$. Therefore, we have obtained the Sturm–Liouville problem

$$-y'' = \lambda y, \quad 0 < x < l; \quad y'(0) = 0, \quad y(l) = 0.$$

We solved this problem in Section 3.4 to obtain eigenvalues and eigenfunctions given by (the reader should review that calculation)

$$\lambda_n = \left(\frac{2n+1}{2}\right)^2 \frac{\pi^2}{l^2}, \quad y_n(x) = \cos\frac{(2n+1)\pi x}{2l}, \quad n = 0, 1, 2 \ldots.$$

By the results in Section 3.4 we know that the eigenfunctions $y_n(x)$ are orthogonal on $0 < x < l$, which means that

$$\int_0^l \cos\frac{(2n+1)\pi x}{2l} \cos\frac{(2m+1)\pi x}{2l}\, dx = 0 \quad \text{for} \quad m \neq n.$$

The solution to the time equation for $g(t)$ is

$$g(t) = ce^{-\lambda k t},$$

and since there are infinitely many values of λ, we have infinitely many such solutions,

$$g_n(t) = c_n e^{-\lambda_n k t}.$$

Summarizing, we have obtained infinitely many product solutions of the form

$$c_n e^{-\lambda_n k t} \cos\frac{(2n+1)\pi x}{2l}, \quad n = 0, 1, 2 \ldots.$$

These functions will solve the PDE (4.17) and the boundary conditions (4.18). Now we superimpose these to form

$$u(x,t) = \sum_{n=0}^{\infty} c_n e^{-\lambda_n k t} \cos\frac{(2n+1)\pi x}{2l}, \tag{4.20}$$

and we properly select the coefficients c_n such that u will satisfy the initial condition (4.19). To this end we have

$$u(x,0) = f(x) = \sum_{n=0}^{\infty} c_n \cos\frac{(2n+1)\pi x}{2l}.$$

This equation is an expansion of the function $f(x)$ in terms of the eigenfunctions $y_n(x)$. We know from Section 3.2 that the coefficients c_n are given by the formula

$$c_n = \frac{1}{\|y_n\|^2}\int_0^l f(x)\cos\frac{(2n+1)\pi x}{2l}\,dx, \quad n = 0, 1, 2 \ldots .$$

A straightforward integration (either by hand or using a computer algebra package) gives

$$\|y_n\|^2 = \int_0^l \cos^2\frac{(2n+1)\pi x}{2l}\,dx = \frac{l}{2}, \quad n = 0, 1, 2 \ldots .$$

Consequently, the solution of (4.17)–(4.19) is given by (4.20), where the coefficients c_n are given by

$$c_n = \frac{2}{l}\int_0^l f(x)\cos\frac{(2n+1)\pi x}{2l}\,dx, \quad n = 0, 1, 2 \ldots .$$

□

We illustrate how to solve a problem with a radiation boundary condition.

EXAMPLE

Consider the diffusion problem

$$u_t = ku_{xx}, \quad 0 < x < 1,\ t > 0, \tag{4.21}$$

$$u(0,t) = 0, \quad u(1,t) + u_x(1,t) = 0, \quad t > 0, \tag{4.22}$$

$$u(x,0) = f(x), \quad 0 < x < 1, \tag{4.23}$$

where the left end is fixed and the right end satisfies a radiation condition. The separation of variables method dictates that we assume $u(x,t) = y(x)g(t)$. Substituting this into the PDE, we obtain, exactly as above,

$$\frac{g'(t)}{kg(t)} = \frac{y''(x)}{y(x)} = -\lambda,$$

where $-\lambda$ is the separation constant. This gives the two ODEs

$$g'(t) = -\lambda k g(t), \quad -y''(x) = \lambda y(x).$$

Substituting the product $u = yg$ into the boundary conditions gives

$$y(0)g(t) = 0, \quad y(1)g(t) + y'(1)g(t) = 0.$$

Since $g(t) \neq 0$, we are forced to take $y(0) = 0$ and $y(1) + y'(1) = 0$, and we are left with the Sturm–Liouville problem

$$-y'' = \lambda y, \quad 0 < x < 1; \quad y(0) = 0, \quad y(1) + y'(1) = 0.$$

Now we determine the eigenvalues, i.e., the values of λ for which this problem has a nontrivial solution. In the case $\lambda = 0$ the solution of the ODE is $y(x) = mx + b$; the two boundary conditions force $m = b = 0$, so zero is not an eigenvalue. To show that there are no negative eigenvalues, we argue as in Section 3.4, using an energy argument. We multiply the ODE by $y(x)$ and integrate from 0 to 1 to get

$$\int_0^1 -yy''dx = \lambda \int_0^1 y^2 dx.$$

On the left side we integrate by parts to obtain

$$-(y(x)y'(x)) \mid_0^1 + \int_0^1 y'^2 dx = \lambda \int_0^1 y^2 dx.$$

Because the integrands are positive, the integral on the right is positive, and the integral on the left is positive. The boundary term is nonnegative since

$$-(y(x)y'(x)) \mid_0^1 = -y(1)y'(1) + y(0)y'(0) = y(1)^2 \geq 0.$$

Therefore, λ cannot be negative.

For $\lambda > 0$ the solution to the ODE is $y(x) = A\cos\sqrt{\lambda}x + B\sin\sqrt{\lambda}x$. The boundary condition $y(0) = 0$ forces $A = 0$, and the boundary condition $y(1) + y'(1) = 0$ forces the condition

$$\sin\sqrt{\lambda} + \sqrt{\lambda}\cos\sqrt{\lambda} = 0,$$

or

$$\sqrt{\lambda} = -\tan\sqrt{\lambda}.$$

This equation, where the variable λ is tied up in a trigonometric function, cannot be solved analytically. But graphs of the functions $\sqrt{\lambda}$ and $-\tan\sqrt{\lambda}$ versus λ (see Figure 4.3) show that the equation has infinitely many positive solutions $\lambda_1, \lambda_2, \ldots$, represented by the intersection points of the two graphs.

These values λ_n are the eigenvalues for the Sturm–Liouville problem, and the corresponding eigenfunctions are

$$y_n = \sin\sqrt{\lambda_n}x, \quad n = 1, 2, \ldots.$$

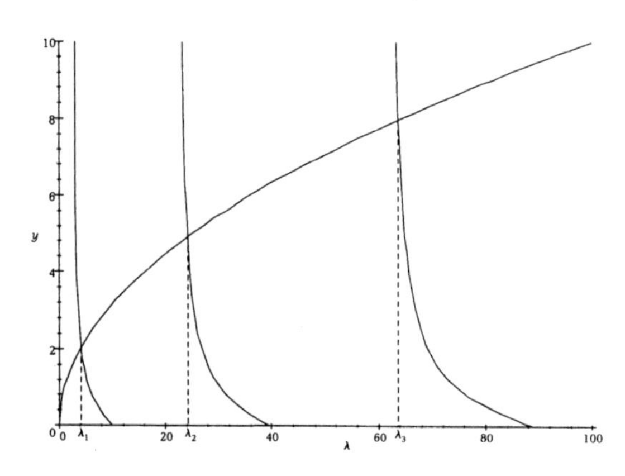

Figure 4.3. Graphical representation of the eigenvalues of the SLP.

Using a solving routine available in computer algebra programs (for example, fsolve in Maple) we obtain the first three eigenvalues: $\lambda_1 = 4.115858, \lambda_2 = 24.13934, \lambda_3 = 63.65911$. By the results in Section 3.4 we know that the eigenfunctions $y_n(x)$ are orthogonal on $0 < x < 1$, which means that

$$\int_0^1 \sin\sqrt{\lambda_n}x \sin\sqrt{\lambda_m}x dx = 0 \quad \text{for} \quad m \neq n.$$

The solution to the time equation is

$$g(t) = ce^{-\lambda kt},$$

and since there are infinitely many values of λ, we have infinitely many such solutions,

$$g_n(t) = c_n e^{-\lambda_n kt}$$

Summarizing, we have obtained infinitely many product solutions of the form

$$c_n e^{-\lambda_n kt} \sin\sqrt{\lambda_n}x, \quad n = 1, 2 \ldots.$$

These functions will solve the PDE (4.21) and the boundary conditions (4.22). Now we superimpose these to form

$$u(x,t) = \sum_{n=1}^{\infty} c_n e^{-\lambda_n kt} \sin\sqrt{\lambda_n}x, \tag{4.24}$$

and we select the coefficients c_n such that u will satisfy the initial condition (4.23). We have

$$u(x,0) = f(x) = \sum_{n=1}^{\infty} c_n \sin\sqrt{\lambda_n}x.$$

This equation is an expansion of the function $f(x)$ in terms of the eigenfunctions $y_n(x)$. We know from Section 3.2 that the coefficients c_n are given by the formula

$$c_n = \frac{1}{\|y_n\|^2}\int_0^1 f(x)\sin\sqrt{\lambda_n}x dx, \quad n = 1, 2 \ldots,$$

or

$$c_n = \frac{1}{\int_0^1 \sin^2\sqrt{\lambda_n}x dx}\int_0^1 f(x)\sin\sqrt{\lambda_n}x dx, \quad n = 1, 2, \ldots. \tag{4.25}$$

Consequently, the solution of (4.21)–(4.23) is given by (4.24), where the coefficients c_n are given by (4.25). □

Computer algebra systems can be extremely helpful in obtaining approximate solutions and their graphs. Let us use Maple to calculate a three-term approximation to the solution in the last example in the case

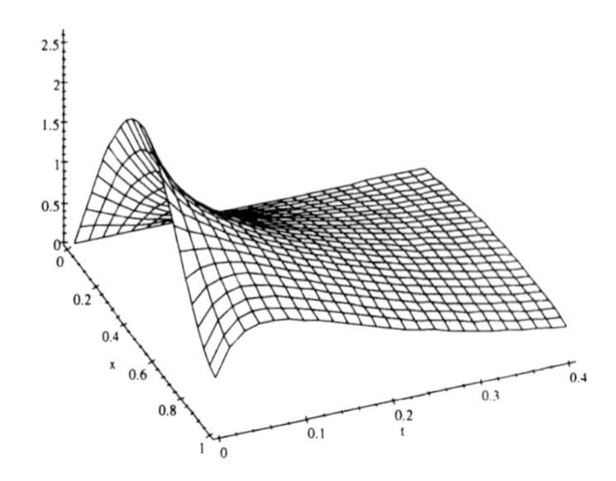

Figure 4.4. Approximate solution surface.

$k = 0.1$ and the initial temperature is $f(x) = 10x(1-x)$. As noted above, the first eigenvalue `lam1` can be determined by the syntax

```
lam1:=fsolve({sqrt (z)+tan(sqrt(z))=0},{z},1..5 ):
```

This gives $lam1 = 4.115858$. In the `fsolve` command the expression 1..5 gives a range of values where a known solution lies. Additional eigenvalues can be obtained in a similar manner. The first three eigenvalues are

```
lam1=4.115858, lam2=24.13934, lam3=63.65911.
```

The coefficient c_1 is calculated via

```
c1:=int(10*x*(1-x)*sin(sqrt(lam1)*x),x=0..1)/
int(sin(sqrt(lam1)*x)^2,x=0..1);
```

The coefficients $c2$ and $c3$ are calculated similarly using $lam2$ and $lam3$. We obtain

$$c1 = 2.13285,\ c2 = 1.040488,\ c3 = -0.219788.$$

Then a three-term approximate solution is

$$\begin{aligned} u(x,t) &\approx 2.13285e^{-4.115858t}\sin\sqrt{4.115858}x \\ &+ 1.040488e^{-24.13934t}\sin\sqrt{24.13934}x \\ &- 0.219788e^{-63.65911t}\sin\sqrt{63.65911}x. \end{aligned}$$

The approximate solution surface is plotted in Figure 4.4. Notice that the it does not match the initial condition accurately near $x = 1$.

Exercises

1. Find an infinite series representation for the solution to the wave problem

$$u_{tt} = c^2 u_{xx}, \quad 0 < x < l,\ t > 0,$$

$$u(0,t) = u_x(l,t) = 0, \quad t > 0,$$

$$u(x,0) = f(x), \quad u_t(x,0) = 0 \quad 0 < x < l.$$

Interpret this problem in the context of waves on a string.

2. Find an infinite-series representation for the solution to the equilibrium problem

$$u_{xx} + u_{yy} = 0, \quad 0 < x < a,\ 0 < y < b,$$

$$u_x(0,y) = u_x(a,y) = 0, \quad 0 < y < b,$$

$$u(x,0) = f(x), \quad u(x,b) = 0, \quad 0 < x < a.$$

Interpret this problem in the context of steady heat flow.

3. Find an infinite-series representation for the solution to the heat absorption–radiation problem

$$u_t = u_{xx}, \quad 0 < x < 1,\ t > 0,$$

$$u_x(0,t) - a_0 u(0,t) = 0, \quad u_x(1,t) + a_1 u(1,t) = 0, \quad t > 0,$$

$$u(x,0) = f(x), \quad 0 < x < 1,$$

where $a_0 < 0$, $a_1 > 0$, and $a_0 + a_1 > -a_0 a_1$. State why there is radiation at $x = 1$, absorption at $x = 0$, and the radiation greatly exceeds the absorption. Choose $a_0 = -0.25$ and $a_1 = 4$ and find the first four eigenvalues; if $f(x) = x(1-x)$, find an approximate solution to the problem and graph the approximate solution surface.

4. Consider a large, circular, tubular ring of circumference $2l$ that contains a chemical of concentration $c(x,t)$ dissolved in water. Let x be the arc-length parameter with $0 < x < 2l$. See Figure 4.5. If the concentration of the chemical is initially given by $c_0(x)$, then $c(x,t)$ satisfies the initial boundary value problem

$$c_t = Dc_{xx}, \quad 0 < x < 2l,\ t > 0,$$

$$c(0,t) = c(2l,t), \quad c_x(0,t) = c_x(2l,t), \quad t > 0,$$

$$c(x,0) = f(x), \quad 0 < x < 2l.$$

These boundary conditions are called *periodic boundary conditions*, and D is

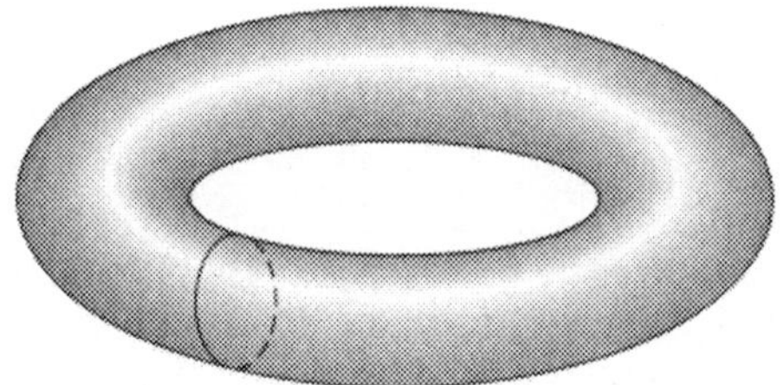

Figure 4.5. Circular ring.

the diffusion constant. Apply the separation of variables method and show that the associated Sturm–Liouville problem has eigenvalues $\lambda_n = (n\pi/l)^2$ for $n = 0, 1, 2, \ldots$ and eigenfunctions $y_0(x) = 1, y_n(x) = A_n \cos(n\pi x/l) + B_n \sin(n\pi x/l)$ for $n = 1, 2, \ldots$. Show that the concentration is given by

$$c(x,t) = \frac{A_0}{2} + \sum_{n=0}^{\infty}(A_n \cos(n\pi x/l) + B_n \sin(n\pi x/l))e^{-n^2\pi^2 Dt/l^2}$$

and find formulae for the A_n and B_n.

4.3 Laplace's Equation

Now we investigate some problems associated with Laplace's equation on a bounded domain. In the first two sections of this chapter we have presented a general method, the Fourier method, to solve Laplace's equation on rectangles. Now we examine circular domains. Moreover, we shall study some general properties of harmonic functions (i.e., solutions to $\Delta u = 0$) in both two and three dimensions.

Suppose we know the temperature on the boundary of a circular, laminar plate of radius R. Can we find the equilibrium temperature inside the plate? We will now find a particularly nice formula for the solution to this problem.

When the region of interest is a circle, we might guess that polar coordinates are more appropriate than rectangular coordinates, so let us formulate the steady-state heat flow problem in polar coordinates r, θ, where $x = r\cos\theta$ and $y = r\sin\theta$. Then a circular plate of radius R can be represented simply as $r \le R$ with $0 \le \theta \le 2\pi$. The unknown temperature inside the plate is $u = u(r,\theta)$, and the given temperature on the boundary of the plate is $u(R,\theta) = f(\theta)$, where f is a known function. We know from prior discussions (Section 1.8) that u must satisfy Laplace's equation $\Delta u = 0$ inside the plate. Therefore, upon representing the Laplacian Δ in polar coordinates (see Section 1.8), we have the following boundary value problem for $u(r,\theta)$:

$$u_{rr} + \frac{1}{r}u_r + \frac{1}{r^2}u_{\theta\theta} = 0, \quad 0 \le r < R, \ 0 \le \theta \le 2\pi, \tag{4.26}$$

$$u(R,\theta) = f(\theta), \quad 0 \le \theta \le 2\pi. \tag{4.27}$$

Implicit are the periodic boundary conditions

$$u(r,0) = u(r,2\pi), \quad u_\theta(r,0) = u_\theta(r,2\pi). \tag{4.28}$$

The separation of variables method works the same way as it did in rectangular coordinates. We assume the product solution $u(r,\theta) = y(r)g(\theta)$ and substitute into the PDE and boundary conditions. When the expression is substituted into the PDE (4.26), we obtain

$$y''(r)g(\theta) + \frac{1}{r}y'(r)g(\theta) + \frac{1}{r^2}y(r)g''(\theta) = 0.$$

This can be written

$$-\frac{r^2y''(r) + ry'(r)}{y(r)} = \frac{g''(\theta)}{g(\theta)} = -\lambda,$$

where $-\lambda$ is the separation constant. Therefore, we obtain two ordinary differential equations for y and g, namely,

$$r^2y''(r) + ry'(r) = \lambda y(r), \quad g''(\theta) = -\lambda g(\theta).$$

The periodic boundary conditions force $g(0) = g(2\pi)$ and $g'(0) = g'(2\pi)$. Therefore, we have the following problem for $g(\theta)$:

$$g''(\theta) = -\lambda g(\theta); \quad g(0) = g(2\pi), \quad g'(0) = g'(2\pi). \tag{4.29}$$

The boundary conditions, which are called *periodic*, are not quite the type that we defined for a Sturm–Liouville problem. But many of the same results are true.

First, it is clear that $\lambda = 0$ is an eigenvalue with corresponding eigenfunction $g_0(r) = 1$. Moreover, there are no negative eigenvalues; if λ is negative, then the ODE has exponential solutions, and exponential solutions cannot satisfy periodicity conditions (or an energy argument could be used). Therefore, let us assume $\lambda = p^2 > 0$. The differential equation for g has general solution

$$g(\theta) = a\cos p\theta + b\sin p\theta.$$

The boundary conditions force

$$(\cos 2\pi p - 1)a + (\sin 2\pi p)b = 0,$$

$$(\sin 2\pi p)a + (1 - \cos 2\pi p)b = 0.$$

This is a system of two linear homogeneous equations for a and b. From a basic result in matrix algebra, we know that it has a nontrivial solution if the determinant of the coefficients is zero, i.e.,

$$(\cos 2\pi p - 1)(1 - \cos 2\pi p) - \sin^2 2\pi p = 0,$$

or, simplifying,

$$\cos 2\pi p = 1.$$

This means that $p = \sqrt{\lambda}$ must be a positive integer, that is,

$$\lambda = \lambda_n = n^2, \quad n = 1, 2, \ldots.$$

Along with $\lambda_0 = 0$, these are the eigenvalues of the problem (4.29). The eigenfunctions are

$$g_0(\theta) = 1, \quad g_n(\theta) = a_n\cos n\theta + b_n\sin n\theta, \quad n = 1, 2, \ldots.$$

Now let us solve the y-equation. Of course, we want bounded solutions. For $\lambda = 0$ the only bounded solution is $y_0(r) = 1$ (the other solution in this case is $\ln r$, which is unbounded). For $\lambda = n^2$ the equation is

$$r^2y''(r) + ry'(r) - n^2y(r) = 0,$$

which is a Cauchy–Euler equation (see the Appendix) with general solution

$$y_n(r) = c_n r^{-n} + d_n r^n. \tag{4.30}$$

We can set $c_n = 0$ because again we seek bounded solutions for $0 \le r \le R$. Thus, setting $d_n = 1$ for all n, we have

$$y_n(r) = r^n, \quad n = 1, 2, \ldots.$$

In summary, we have constructed solutions of the given boundary value problem of the form $u_0(r, \theta) = \text{constant} = a_0/2$, $u = u_n(r, \theta) = r^n(a_n \cos n\theta + b_n \sin n\theta)$, $n = 1, 2, \ldots$. Now, to satisfy the boundary condition at $r = R$ we form the linear combination

$$u(r, \theta) = \frac{a_0}{2} + \sum_{n=1}^{\infty} r^n(a_n \cos n\theta + b_n \sin n\theta). \tag{4.31}$$

The boundary condition $u(R, \theta) = f(\theta)$ then yields

$$f(\theta) = \frac{a_0}{2} + \sum_{n=1}^{\infty} R^n(a_n \cos n\theta + b_n \sin n\theta),$$

which is the full-range Fourier series for $f(\theta)$ (see Section 3.3). Therefore, the coefficients are given by

$$a_n = \frac{1}{\pi R^n}\int_0^{2\pi} f(\theta)\cos n\theta \, d\theta, \quad n = 0, 1, 2, \ldots, \tag{4.32}$$

$$b_n = \frac{1}{\pi R^n}\int_0^{2\pi} f(\theta)\sin n\theta \, d\theta, \quad n = 1, 2, \ldots. \tag{4.33}$$

So the solution to the BVP (4.26)–(4.28) is given by (4.31) with the coefficients given by (4.32)–(4.33).

As it turns out, this series solution can be manipulated to obtain a simple formula, called Poisson's integral formula. Let us substitute the coefficients given in formulae (4.32) and (4.33) (after changing the dummy integration variable from θ to ϕ) into the solution formula (4.31) to obtain

$$\begin{aligned} u(r, \theta) = {} & \frac{1}{2\pi}\int_0^{2\pi} f(\phi)d\phi \\ & + \sum_{n=1}^{\infty} \frac{r^n}{\pi R^n}\int_0^{2\pi} f(\phi)(\cos n\phi \cos n\theta + \sin n\phi \sin n\theta)d\phi \\ = {} & \frac{1}{2\pi}\int_0^{2\pi} f(\phi)\left(1 + 2\sum_{n=1}^{\infty}\left(\frac{r}{R}\right)^n \cos n(\theta - \phi)\right)d\phi. \end{aligned}$$

But the infinite sum in the integrand can be determined exactly as follows. Recalling the identity $\cos\alpha = \frac{1}{2}(e^{i\alpha} + e^{-i\alpha})$, we can write

$$1 + 2\sum_{n=1}^{\infty}\left(\frac{r}{R}\right)^n \cos n(\theta - \phi) = 1 + \sum_{n=1}^{\infty}\left(\frac{r}{R}\right)^n e^{in(\theta-\phi)} + \sum_{n=1}^{\infty}\left(\frac{r}{R}\right)^n e^{-in(\theta-\phi)}.$$

But each series on the right side is a geometric series, and we know from calculus that

$$\sum_{n=1}^{\infty} z^n = \frac{z}{1-z}, \quad \text{provided that } |z| < 1.$$

Using this result, we get

$$1 + \sum_{n=1}^{\infty}\left(\frac{r}{R}\right)^n \cos n(\theta - \phi) = 1 + \frac{re^{i(\theta-\phi)}}{R - re^{i(\theta-\phi)}} + \frac{re^{-i(\theta-\phi)}}{R - re^{-i(\theta-\phi)}}$$

$$= \frac{R^2 - r^2}{R^2 + r^2 - 2rR\cos(\theta - \phi)}.$$

So we have the formula

$$u(r,\theta) = \frac{1}{2\pi}\int_0^{2\pi} \frac{(R^2 - r^2)f(\phi)}{R^2 + r^2 - 2rR\cos(\theta - \phi)}\,d\phi, \tag{4.34}$$

which is *Poisson's integral formula* for the solution to the Dirichlet problem (4.26)–(4.27).

We have seen in the last section and in the previous discussion how to use separation of variables to derive solutions to the two-dimensional Laplace's equation on special bounded domains: rectangles and circles. The calculation is similar for portions of circles (wedges) and for annular domains. For other types of domains other methods are used.

The most common method for solving Laplace's equation on any bounded domain is to use a numerical method or computer method. For example, in the method of finite differences (see Exercise 6 in Section 1.8 and Section 4.8) the derivatives are approximated by difference quotients, and the solution $u(x, y)$ is determined at discrete lattice points in the region. There is another popular method, the finite element method, that again yields the solution at discrete points. The finite difference method is discussed further in Section 4.8. Students familiar with complex variable theory will recall that some domains can be mapped conformally onto a simple domain where Laplace's equation can be solved; conformal transformations preserve solutions to Laplace's equation, and so this method can result in a closed form solution, i.e., a formula, for the answer. Most elementary texts on complex analysis discuss this method (see, for example, Churchill (1960)).

Here we want to introduce some of the basic properties of solutions to Laplace's equation on bounded domains. Many of the results follow from the *divergence theorem*, which states that

$$\int_{\Omega} \operatorname{div}\phi \, dV = \int_{\partial\Omega} \phi \cdot n \, dA. \tag{4.35}$$

The divergence theorem is a version of the fundamental theorem of calculus in three dimensions—it converts the integral of derivatives, a divergence, to an integral over the boundary. We recall from elementary calculus that the divergence measures the local outflow per unit volume of a vector field ϕ, so the divergence theorem states that the net outflow, per unit volume, in a region Ω must equal the flux through the boundary (surface) $\partial\Omega$ of that region. Recall that the integral on the right is a flux integral through the surface oriented by the outward unit normal vector n. Here we are taking a fluid flow interpretation in speaking of outflow, but the interpretation is valid for any smooth vector field. We always take Ω to be a nice region with a smooth boundary or with a boundary made up of finitely many smooth sections. There are some pathological domains on which the divergence theorem does not apply, but we do not consider those. Moreover, we always assume that the functions involved in our discussion are sufficiently smooth for the theorems to apply. For example, the functions should have two continuous partial derivatives in Ω, be continuous on $\Omega \cup \partial\Omega$, and have first partial derivatives that extend continuously to the boundary.

Two important integral identities, called Green's identities, follow from the divergence theorem. To derive these identities we also need the vector identity

$$\operatorname{div}(u\phi) = u \operatorname{div}\phi + \phi \cdot \operatorname{grad} u$$

(see Exercise 4, Section 1.7). Integrating this identity over the volume Ω and using the divergence theorem gives

$$\int_{\partial\Omega} u\phi \cdot n \, dA = \int_{\Omega} u \operatorname{div}\phi \, dV + \int_{\Omega} \phi \cdot \operatorname{grad} u \, dV.$$

Setting $\phi = \operatorname{grad} v$ for a scalar function v then gives

$$\int_{\partial\Omega} u \operatorname{grad} v \cdot n \, dA = \int_{\Omega} u\Delta v dV + \int_{\Omega} \operatorname{grad} v \cdot \operatorname{grad} u \, dV. \tag{4.36}$$

Here we have used the fact that $\operatorname{div}(\operatorname{grad} u) = \Delta u$. In particular, if we set $v = u$, we obtain

$$\int_{\partial\Omega} u \operatorname{grad} u \cdot n \, dA = \int_{\Omega} u\Delta u \, dV + \int_{\Omega} \operatorname{grad} u \cdot \operatorname{grad} u \, dV, \tag{4.37}$$

and this is *Green's first identity*. To obtain Green's second identity, interchange v and u in equation (4.36) and subtract that result from (4.36) to get

$$\int_{\Omega} u\Delta v \, dV = \int_{\Omega} v\Delta u \, dV + \int_{\partial\Omega} (u \operatorname{grad} v - v \operatorname{grad} u) \cdot n \, dA. \tag{4.38}$$

This is *Green's second identity*; it can be regarded as an integration by parts formula for the Laplacian Δ; note how the Laplacian is taken off v and put on u and a boundary term is produced.

Let us show how to use Green's identities to prove interesting facts about problems involving Laplace's equation. We show that solutions to the Dirichlet problem in $\mathbf{R}^3$ are unique (it is, of course, the same proof in $\mathbf{R}^2$). Thus, consider the problem

$$\Delta u = 0, \quad \text{in } \Omega,$$

$$u = f(x, y, z), \quad \text{on } \partial\Omega.$$

By way of contradiction, assume that there are two solutions u and v. Then it easily follows that the difference $w \equiv u - v$ must satisfy the problem

$$\Delta w = 0, \quad \text{in } \Omega,$$

$$w = 0, \quad \text{on } \partial\Omega.$$

Now multiply this PDE by w and integrate over the region Ω to get

$$\int_\Omega w \Delta w dV = 0.$$

But now we can use Green's first identity (4.37) to get

$$\int_{\partial\Omega} w \operatorname{grad} w \cdot n \, dA - \int_\Omega \operatorname{grad} w \cdot \operatorname{grad} w \, dV = 0.$$

The first integral on the left is zero because $w = 0$ on the boundary. Therefore,

$$\int_\Omega \operatorname{grad} w \cdot \operatorname{grad} w \, dV = 0.$$

Since the integrand is never negative, we must have $\operatorname{grad} w = 0$, or $w =$ constant in Ω. Since w is continuous and is zero on the boundary, the constant must be zero, and so $w = 0$ in $\Omega \cup \partial\Omega$. This means that $u = v$ in $\Omega \cup \partial\Omega$. So there cannot be two solutions. The reader should recognize the proof above as an "energy argument" of the type introduced in Section 3.4 for SLPs.

Another interesting property of harmonic functions that satisfy Dirichlet-type boundary conditions is that they minimize the *energy integral*

$$E(w) \equiv \int_\Omega \operatorname{grad} w \cdot \operatorname{grad} w \, dV.$$

In electrostatics, for example, where w is the electric field potential, the energy E is the energy stored in the electrostatic field. This principle is called *Dirichlet's principle*: Suppose u satisfies

$$\Delta u = 0 \quad \text{in } \Omega; \qquad u = f \quad \text{on } \partial\Omega.$$

Then $E(u) \le E(w)$ for all w satisfying $w = f$ on $\partial\Omega$. In other words, of all functions that satisfy the boundary condition, the solution to Laplace's equation is the one that minimizes the energy.

It is not hard to prove Dirichlet's principle. Let $w = u + v$ where $v = 0$ on the boundary $\partial\Omega$. Then

$$\begin{aligned}
E(w) &= E(u+v) \\
&= \int_\Omega \operatorname{grad}(u+v)\cdot\operatorname{grad}(u+v)dV \\
&= \int_\Omega \operatorname{grad} u \cdot \operatorname{grad} u \; dV + 2\int_\Omega \operatorname{grad} u \cdot \operatorname{grad} v \; dV \\
&\quad + \int_\Omega \operatorname{grad} v \cdot \operatorname{grad} v \; dV \\
&= E(u) + E(v) + 2\int_\Omega \operatorname{grad} u \cdot \operatorname{grad} v \; dV.
\end{aligned}$$

But the integral term on the right is zero by equation (4.36). And because $E(v) \ge 0$, we have $E(u) \le E(w)$, which completes the proof.

Exercises

1. Show that the solution $u = u(r,\theta)$ to the *exterior* boundary value problem

$$\Delta u = 0, \quad r > R, 0 \le \theta < 2\pi,$$

$$u(R,\theta) = f(\theta), 0 \le \theta < 2\pi,$$

is given by

$$u(r,\theta) = \frac{1}{2\pi}\int_0^{2\pi} \frac{(r^2 - R^2)f(\phi)}{R^2 + r^2 - 2rR\cos(\theta - \phi)}\, d\phi.$$

(Follow the example in the book but choose the other bounded solution in equation (4.30).)

2. Use separation of variables to find an infinite series solution to the following boundary value problem on a wedge:

$$\Delta u = 0, \quad r < R, \quad 0 \le \theta < \pi/2,$$

$$u(R,\theta) = f(\theta), \quad 0 < \theta < \pi/2,$$

$$u(r,0) = 0, \quad u(r,\pi/2) = 0, \quad 0 < r < R.$$

3. Solve the boundary value problem in Exercise 2 with the boundary conditions along the edges $\theta = 0, \ \theta = \pi/2$ replaced by

$$u(r,0) = 0, \quad u_\theta(r,\pi/2) = 0, \quad 0 < r < R.$$

4. Use Poisson's integral formula to prove that the value of the equilibrium temperature at the center of a disk is the average of the temperatures on the boundary.

5. Prove Dirichlet's principle for the Neumann problem: Let

$$E(w) \equiv \frac{1}{2}\int_{\Omega} \operatorname{grad} w \cdot \operatorname{grad} w \, dV - \int_{\partial\Omega} hw \, dA.$$

If u is the solution to the BVP

$$\Delta u = 0, \quad \text{in} \quad \Omega,$$

$$n \cdot \operatorname{grad} u = h \quad \text{on} \ \partial\Omega,$$

then $E(u) \leq E(w)$ for all w sufficiently smooth. Observe that the average value of h on the boundary is zero.

6. In three dimensions consider the radiation boundary value problem

$$\Delta u - cu = 0, \quad \text{in} \quad \Omega,$$

$$n \cdot \operatorname{grad} u + au = 0, \quad \text{on} \ \partial\Omega,$$

where $a, c > 0$ and Ω is a bounded region. Show that the only solution is $u = 0$. (Hint: Use an energy argument. Multiply the PDE by u, integrate, and then use Green's first identity.) Use this result to show that solutions to the boundary value problem

$$\Delta u - cu = g(x, y, z), \quad (x, y, z) \in \Omega,$$

$$n \cdot \operatorname{grad} u + au = f(x, y, z), \quad \text{on} \ \partial\Omega,$$

must be unique.

7. Solve the boundary value problem

$$\Delta u = 0, \quad r < R, \quad 0 \leq \theta < 2\pi,$$

$$u(R, \theta) = 4 + 3\sin\theta, \quad 0 \leq \theta < 2\pi.$$

8. Suppose $u = u(x, y, z)$ satisfies the Neumann problem

$$\Delta u = 0, \quad \text{in} \ \Omega,$$

$$n \cdot \operatorname{grad} u = 0 \quad \text{on} \ \partial\Omega.$$

Show that u must be constant on Ω.

4.4 Cooling of a Sphere

We have observed that the separation of variables method can be applied to several types of problems on bounded spatial domains. The problem must be linear and have homogeneous boundary conditions. In this section we present the solution to a classical problem in three dimensions, the cooling of a sphere. The assumed symmetries in the problem will permit us to reduce the dimension of the problem to one spatial dimension and time, and we can follow the same procedures as in previous sections.

The problem is this: Given a sphere whose initial temperature depends only on the distance from the center (e.g., a constant initial temperature) and whose boundary is kept at a constant temperature, predict the temperature at any point inside the sphere at a later time. This is the problem, for example, of determining the temperature of the center of a potato that has been put in a hot oven. The reader might also conjecture that this problem is important for medical examiners who want to determine the time of an individual's death. Early researchers, notably Kelvin, used this problem to determine the age of the earth based on assumptions about its initial temperature and its temperature today.

This cooling problem may remind the reader of Newton's law of cooling, which is encountered in ordinary differential equations texts. Recall that this law states that the rate at which a body cools is proportional to the difference of its temperature and the temperature of the environment. Quantitatively, if $T = T(t)$ is the temperature of a body and T_e is the temperature of its environment, then $T'(t) = K(T_e - T)$, where K is the constant of proportionality. But the reader should note that this law applies only in the case that the body has a uniform, homogeneous temperature. In the PDE problem we are considering, the temperature may vary radially throughout the body.

For simplicity, let us consider a sphere of radius $\rho = \pi$ whose initial temperature is $T_0 =$ constant. We will assume that the boundary is held at zero degrees for all time $t > 0$. If u is the temperature, then in general u will depend on three spatial coordinates and time. But a little reflection shows that the temperature will depend only on the distance from the center of the sphere and on time. Evidently, the temperature u must satisfy the heat equation

$$u_t = k\Delta u,$$

where k is the diffusivity and Δ is the Laplacian (see Section 1.7). It should be clear that a spherical coordinate system ρ, ϕ, θ is more appropriate than a rectangular system, and $u = u(\rho, t)$. Because u does not depend on the angles ϕ and θ, the Laplacian takes on a particularly simple form:

$$\Delta = \frac{\partial^2}{\partial \rho^2} + \frac{2}{\rho}\frac{\partial}{\partial \rho}$$

(see Section 1.8). Therefore, we may formulate the model as

$$u_t = k(u_{\rho\rho} + \frac{2}{\rho}u_\rho), \quad 0 \le \rho < \pi, \ t > 0, \tag{4.39}$$

$$u(\pi, t) = 0, \quad t > 0, \tag{4.40}$$

$$u(\rho, 0) = T_0, \quad 0 \le \rho < \pi. \tag{4.41}$$

Observe that there is an implied boundary condition at $\rho = 0$, namely that the temperature should remain bounded.

To solve problem (4.39)–(4.41) we assume $u(\rho, t) = y(\rho)g(t)$. Substituting into the PDE and separating variables gives, in the usual way,

$$\frac{g'(t)}{kg(t)} = \frac{y''(\rho) + \frac{2}{\rho}y'(\rho)}{y(\rho)} = -\lambda,$$

where $-\lambda$ is the separation constant. Substituting $u(\rho, t) = y(\rho)g(t)$ into the boundary condition (4.40) gives $y(\pi) = 0$, and the boundedness of u at $\rho = 0$ forces $y(0)$ to be bounded. Therefore, we have the boundary value problem

$$-y''(\rho) - \frac{2}{\rho}y'(\rho) = \lambda y(\rho), \quad 0 \leq \rho < \pi, \tag{4.42}$$

$$y(0) \text{ bounded}, \quad y(\pi) = 0. \tag{4.43}$$

Note that this is a singular SLP. If $\lambda = 0$, then the equation has the form $-y'' - \frac{2}{\rho}y' = 0$, which has general solution $y = a/\rho + b$; boundedness at zero implies $a = 0$, and $y(\pi) = 0$ implies $b = 0$. So $\lambda = 0$ is not an eigenvalue. If $\lambda \neq 0$, it may not be clear to the reader how to solve the variable-coefficient equation (4.42). However, if we introduce the new function $Y(\rho)$ defined by

$$Y(\rho) = \rho y(\rho),$$

then the differential equation (4.42) becomes the familiar equation

$$-Y''(\rho) = \lambda Y(\rho)$$

(guessing these types of transformations of the dependent variable is not always easy or even possible). If $\lambda > 0$, then $Y = a \cos \sqrt{\lambda}\rho + b \sin \sqrt{\lambda}\rho$, which gives

$$y(\rho) = \frac{1}{\rho}(a \cos \sqrt{\lambda}\rho + b \sin \sqrt{\lambda}\rho).$$

We must have $a = 0$, since $\cos \sqrt{\lambda}\rho/\rho$ is unbounded at $\rho = 0$. Therefore, applying the boundary condition $y(\pi) = 0$ then gives

$$b \sin \sqrt{\lambda}\pi = 0,$$

which in turn gives the eigenvalues

$$\lambda = \lambda_n = n^2, \quad n = 1, 2, \ldots.$$

The eigenfunctions are

$$y_n(\rho) = \frac{\sin n\rho}{\rho}, \quad n = 1, 2, \ldots$$

We leave it as an exercise to show that there are no negative eigenvalues (see Exercise 1).

The corresponding solutions to the time equation for $g(t)$ are easily found to be $g_n(t) = c_n e^{-n^2kt}$. Thus we have determined infinitely many solutions of the form

$$c_n e^{-n^2kt}\frac{\sin n\rho}{\rho}, \quad n = 1, 2, \ldots,$$

that satisfy the PDE and the boundary conditions. To satisfy the initial condition (4.41) we form the linear combination

$$u(\rho, t) = \sum_{n=1}^{\infty} c_n e^{-n^2 kt} \frac{\sin n\rho}{\rho}.$$

Applying the initial condition, we obtain

$$u(\rho, 0) = T_0 = \sum_{n=1}^{\infty} c_n \frac{\sin n\rho}{\rho},$$

or

$$T_0\rho = \sum_{n=1}^{\infty} c_n \sin n\rho.$$

The right side of this equation is the Fourier sine series of the function $T_0\rho$ on the interval $[0, \pi]$. Therefore, the Fourier coefficients c_n are given by

$$c_n = \frac{2}{\pi} \int_0^{\pi} T_0\rho \sin n\rho \, d\rho.$$

The integral can be calculated by hand (using integration by parts) or by a computer algebra system. In any case, we get

$$c_n = (-1)^{n+1} \frac{2T_0}{n}.$$

Consequently, we have derived the solution formula

$$u(\rho, t) = \sum_{n=1}^{\infty} (-1)^{n+1} \frac{2T_0}{n} e^{-n^2 kt} \frac{\sin n\rho}{\rho}. \tag{4.44}$$

This formula gives the temperature at time t at a distance ρ from the center of the sphere.

Exercises

1. Show that the eigenvalue problem (4.42)–(4.43) has no negative eigenvalues.
2. In the cooling of the sphere problem take $R = \pi$ inches, $T_0 = 37$ degrees Celsius, and $k = 5.58$ inches-squared per hour. Using (4.44), plot the temperature at the center of the sphere as a function of time; take $t = 0$ to $t = 1$ hour.
3. This problem deals with the cooling of a sphere with a radiation boundary condition. A spherical ball of radius R has a diffusivity k and initial temperature $u = f(\rho)$, $0 \le \rho \le R$, which depends only on the distance ρ from the center. Heat radiates from the surface of the sphere via the law

$$u_\rho(R, t) = -hu(R, t),$$

where h is a positive constant and $Rh < 1$. Find a formula for the temperature $u(\rho, t)$ in the sphere. Hint: Find the eigenvalues λ_n as the positive roots of the equation

$$\tan R\lambda = \frac{R\lambda}{1 - Rh}.$$

Next, take $R = \pi$ inches, $k = 5.58$ inches-squared per hour, $f(\rho) = 37$ degrees Celsius, and $h = 0.1$ per inch, and calculate a four term approximate solution to the temperature function.

4. Consider a sphere of unit radius on which there is a prescribed potential u depending only on the spherical coordinate angle ϕ (see Section 1.8). This exercise deals with using separation of variables to find an expression for the bounded potential that satisfies Laplace's equation inside the sphere. The boundary value problem is

$$\Delta u = 0, \quad 0 \le \rho < 1, \ 0 < \phi < \pi; \quad u(1, \phi) = f(\phi), \quad 0 \le \phi \le \pi,$$

where $u = u(\rho, \phi)$. Notice that u will not depend on the polar angle θ because of the symmetry in the boundary condition.

(a) Assume $u = R(\rho)Y(\phi)$ and derive the two equations

$$-((1 - x^2)y')' = \lambda y, \quad -1 < x < 1; \quad (\rho^2 R')' - \lambda R = 0,$$

where $x = \cos\phi$ and $y(x) = Y(\arccos x)$.

(b) The y-equation in part (a) is *Legendre's differential equation*, and it has bounded, continuous solutions on $[-1, 1]$ when

$$\lambda = \lambda_n = n(n+1), \quad y = y_n(x) = P_n(x), \quad n = 0, 1, 2, \ldots,$$

where the P_n are polynomial functions called the *Legendre polynomials*. The Legendre polynomials are orthogonal on $[-1, 1]$, and they are given by Rodrigues' formula

$$P_n(x) = \frac{1}{n!2^n} \frac{d^n}{dx^n}[(x^2 - 1)^n].$$

Go through a formal calculation and derive a formula for the solution to the given BVP in the form

$$u(\rho, \phi) = \sum_{n=0}^{\infty} c_n \rho^n P_n(\cos\phi),$$

where

$$c_n = \frac{1}{||P_n||^2} \int_0^{\pi} f(\phi) P_n(\cos\phi) \sin\phi \, d\phi.$$

(c) Find the first four Legendre polynomials using Rodrigues' formula.

(d) If $f(\phi) = \sin\phi$, find an approximate solution.

5. Estimate the age of the earth. Use the data in Exercise 5 of Section 2.4.

4.5 Diffusion in a Disk

In the last section we solved a diffusion problem in spherical coordinates and found, provided that we assumed angular symmetry, that it was a routine application of the separation of variables method. Now we will solve a diffusion problem in cylindrical or polar coordinates, and we will find that the task is not so routine because we will obtain an equation (Bessel's equation) that may not be familiar to the reader. Therefore, we will use a software package to help us understand some of the steps along the way. (The instructor may find that this problem is well suited for presentation in a computer laboratory setting).

Consider a circular, planar disk of radius R whose initial temperature is a function of the radius alone and whose boundary is held at zero degrees. We are interested in how heat diffuses throughout the plate.

Our intuition dictates that the temperature u in the disk depends only on time and the distance r from the center, that is, $u = u(r,t)$. We may make this assumption because there is nothing in the initial condition or boundary condition to cause heat to diffuse in an angular direction—heat will flow only along rays emanating from the origin. We know that u must satisfy the two-dimensional heat equation $u_t = k\Delta u$ in $0 \le r < R$, where k is the diffusivity and Δ is the Laplacian. Here, it is clear that we should write the Laplacian in polar coordinates, and because there is no θ dependence, the heat equation is simply (see Section 1.8 for the formula for the Laplacian in polar coordinates)

$$u_t = k(u_{rr} + \frac{1}{r}u_r), \quad 0 \le r < R,\ t > 0. \tag{4.45}$$

The boundary condition is

$$u(R,t) = 0, \quad t > 0, \tag{4.46}$$

and the initial condition is

$$u(r,0) = f(r), \quad 0 \le r < R, \tag{4.47}$$

where f is a given radial temperature distribution. It is understood that the temperature should be bounded at $r = 0$. Equations (4.45)–(4.47) form the PDE model.

We proceed by separating variables. Taking $u(r,t) = y(r)g(t)$, the PDE splits into

$$\frac{g'(t)}{kg(t)} = \frac{y''(r) + \frac{1}{r}y'(r)}{y(r)} = -\lambda.$$

The equation for $g(t)$ is solved in the usual way to obtain

$$g(t) = e^{-\lambda kt}.$$

The y-equation is

$$y''(r) + \frac{1}{r}y'(r) = -\lambda y(r). \tag{4.48}$$

Upon multiplying by r, we can rewrite this equation as

$$-(ry'(r))' = \lambda r y(r). \tag{4.49}$$

Condition (4.46) leads to the boundary condition

$$y(R) = 0, \tag{4.50}$$

and we impose the boundedness requirement

$$y(0) \quad \text{bounded}. \tag{4.51}$$

Note that the ODE (4.49) along with the boundary condition is a singular Sturm–Liouville problem. Nevertheless, it is an eigenvalue problem, and we need to solve it.

It is easily seen that $\lambda = 0$ is not an eigenvalue (in this case the general solution of (4.49) is $y = a + b \ln r$ and $b = 0$ by boundedness and $a = 0$ by the condition $y(R) = 0$). We leave it as an exericise to show that there are no negative eigenvalues (Exercise 1). So let us consider the case where λ is positive. The task is to solve (4.48) or (4.49). In the last section we found a simple substitution for the similar equation $y'' + (2/\rho)y' = \lambda y$ that transformed it to a familiar equation; but (4.48) is slightly different (by just a factor of 2 in one term), and a simple change of variables does not work. Another strategy is to ask the computer whether a solution is possible. Using `dsolve` in Maple, we type

```
eqn:= diff(y(r),r,r)+(1/r)
      *diff(y(r),r)+lambda*y(r)=0;
sol:= dsolve ({eqn},{y(r)});
```

and Maple returns

$$\texttt{sol:=y(r)=\_C1 BesselJ } (0, \sqrt{\lambda} r)$$
$$+ \texttt{ \_C2 BesselY } (0, \sqrt{\lambda} r).$$

Maple found the general solution in terms of two special functions named `BesselJ(0,z)` and `BesselY(0,z)`. These functions are classical *Bessel functions*, denoted in science literature by $J_0(z)$ and $Y_0(z)$, respectively; so the general solution to (4.48), which is a special case of *Bessel's differential equation*, is

$$y(r) = c_1 J_0(\sqrt{\lambda} r) + c_2 Y_0(\sqrt{\lambda} r).$$

If we ask our computer algebra package to plot $Y_0(z)$, we will notice that it blows up at $z = 0$; in fact, it can be shown that $Y_(z)$ behaves like $\ln z$ for small, positive z (see Figure 4.6). Therefore, we discard the second term in the solution to maintain boundedness; so we set $c_2 = 0$. Therefore, we have

$$y(r) = c_1 J_0(\sqrt{\lambda} r). \tag{4.52}$$

A graph of $J_0(z)$ is shown in Figure 4.6.

The Bessel function $J_0(z)$ is a nicely behaved, oscillatory function. It is quite similar to the cosine function, only its oscillations decay as z increases. It has infinitely many zeros z_n, $n = 1, 2, 3, \ldots$; the first few are $z_1 = 2.405$, $z_2 = 5.520$, $z_3 = 8.654$, $z_4 = 11.790$. Because Bessel functions occur in so many physical contexts, many facts about them have been collected in handbooks and treatises on Bessel functions (e.g., see Abramowitz and Stegun (1965)). For example, one can apply the Taylor series method to solve the ODE (4.48) to find the series representation

$$J_0(z) = \sum_{n=0}^{\infty} \frac{(-1)^n}{n! 2^{2n}} z^{2n}.$$

One can also show that for large z,

$$J_0(z) \sim \sqrt{\frac{2}{\pi z}} \cos(z - \frac{\pi}{4}),$$

which shows the "decaying cosine" behavior.

Now we apply the boundary condition (4.50) to (4.52), and we obtain

$$y(R) = c_1 J_0(\sqrt{\lambda} R) = 0.$$

Thus

$$\sqrt{\lambda} R = z_n\,, \quad z = 1, 2, 3, \ldots,$$

where the z_n are the zeros of J_0. Consequently, the eigenvalues are

$$\lambda_n = \frac{z_n^2}{R^2}, \quad n = 1, 2, 3, \ldots,$$

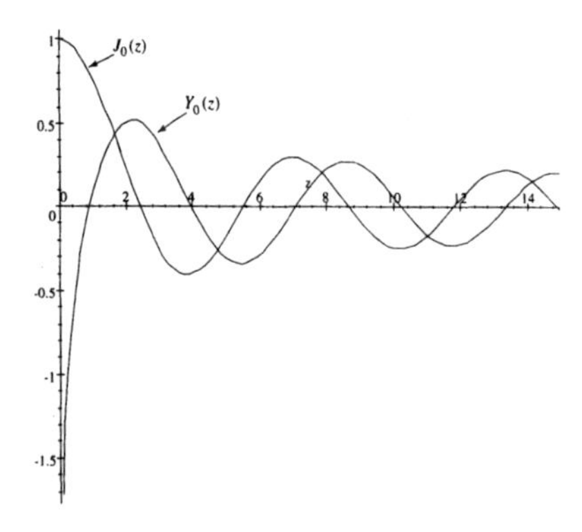

Figure 4.6. Graphs of the Bessel functions $J_0(z)$ and $Y_0(z)$ for $0 \leq z \leq 15$.

and the corresponding eigenfunctions are

$$y_n(r) = J_0(\frac{z_n r}{R}), \quad n = 1, 2, 3, \ldots.$$

In summary, we have constructed solutions $g_n(t)y_n(r)$ of the form

$$e^{-\lambda_n kt} J_0(\frac{z_n r}{R}), \quad n = 1, 2, 3, \ldots,$$

that satisfy the PDE and the boundary conditions.

To satisfy the initial condition (4.47) we form the linear combination

$$u(r, t) = \sum_{n=1}^{\infty} c_n e^{-\lambda_n kt} J_0(\frac{z_n r}{R}) \tag{4.53}$$

and determine the appropriate coefficients c_n. The initial condition forces

$$u(r, 0) = f(r) = \sum_{n=1}^{\infty} c_n J_0(\frac{z_n r}{R}).$$

How do we find the c_n? Orthogonality! This last equation is a Fourier-Bessel expansion, and the Bessel functions $J_0(\frac{z_n r}{R})$ satisfy the orthogonality condition

$$\int_0^R J_0(\frac{z_n r}{R}) J_0(\frac{z_m r}{R}) r dr = 0, \quad n \neq m \tag{4.54}$$

(see Exercise 2). Notice the factor of r in the integrand; we say that r is a weight function and that the Bessel functions $J_0(\frac{z_n r}{R})$ are orthogonal with respect to the weight function r. The orthogonality allows us to determine the coefficients in the standard way: Multiply equation (4.53) by $rJ_0(\frac{z_m r}{R})$ and integrate from $r = 0$ to $r = R$, interchange the summation and the integral, and then use the weighted orthogonality relation to collapse the infinite sum to one term. We obtain

$$\int_0^R f(r) J_0(\frac{z_n r}{R}) r dr = c_m \| J_0(\frac{z_m r}{R}) \|^2,$$

where

$$\| J_0(\frac{z_n r}{R}) \|^2 = \int_0^R J_0(\frac{z_m r}{R})^2 r dr.$$

Therefore, the coefficients are given by

$$c_n = \frac{\int_0^R f(r) J_0(\frac{z_n r}{R}) r dr}{\| J_0(\frac{z_n r}{R}) \|^2}. \tag{4.55}$$

In conclusion, the formal solution to the initial boundary value problem (4.45)–(4.47) is given by (4.53) with coefficients given by (4.55).

In the exercises the reader is asked to calculate this solution numerically in special cases.

We conclude by making some brief remarks about special functions. The Bessel functions J_0 and Y_0 encountered above as solutions of Bessel's equation are just two examples of large classes of Bessel functions in the literature. Problems in cylindrical geometry often lead to such functions. Moreover, other special functions (Legendre polynomials, Laguerre polynomials, and others) occur in a similar way for other problems in other coordinate systems and domains; that is, they occur as solutions to eigenvalue problems for ordinary differential equations that arise from the separation of variables procedure. Generally, special functions arising in this manner possess orthogonality properties that permit the expansion of the solution in terms of those functions.

Exercises

1. Show that the eigenvalue problem (4.49)–(4.51) has no negative eigenvalues. Hint: Use an energy argument—multiply the ODE by y and integrate from $r = 0$ to $r = R$; use integration by parts and use the boundedness at $r = 0$ to get the boundary term to vanish.

2. Derive the weighted orthogonality relation (4.54). Hint: Proceed as in Section 3.4 for regular Sturm–Liouville problems.

3. For the initial boundary value problem (4.45)–(4.47) take $R = 1$, $k = 0.25$, and $f(r) = 5r^3(1 - r)$. Use a computer algebra program to determine a three-term approximation to the solution using (4.53). Sketch a few radial temperature profiles $u(r, t)$ for several fixed values of t.

4. When $R = 1$ sketch the first four eigenfunctions $J_0(\frac{z_n r}{R})$, $n = 1, 2, 3, 4$.

5. (Computer algebra project) Consider the boundary value problem

$$y'' + axy = 0, \quad 0 < x < L,$$

$$y'(0) = 0, \quad y(L) = 0,$$

where a is a positive constant. Find the smallest positive value of L for which the boundary value problem has a nontrivial solution. The answer is that L can be calculated from the equation

$$\frac{2}{3}\sqrt{a}L^{2/3} = 1.86635.$$

Hint: in Maple use `dsolve` to find the general solution as a linear combination of the two Bessel functions $J_{1/3}$ and $Y_{1/3}$; use the left boundary condition in limiting form to determine one constant in terms of the other, and then use the right boundary condition and `fsolve` to determine the appropriate value of L.

4.6 Sources on Bounded Domains

Now we show how to solve problems on bounded domains when sources are present. A source term represents an outside influence in the system and leads to inhomogeneity in the PDE. We will demonstrate two methods for solving PDEs with sources, Duhamel's principle and eigenfunction expansions. For illustration we consider a simple diffusion model given by

$$u_t - ku_{xx} = f(x, t), \quad 0 < x < \pi, \ t > 0, \tag{4.56}$$

$$u(0, t) = u(\pi, t) = 0, \quad t > 0, \tag{4.57}$$

$$u(x, 0) = 0, \quad 0 < x < \pi. \tag{4.58}$$

Interpreted in the context of heat flow, $u = u(x, t)$ is the temperature of a rod whose initial temperature is zero and whose ends are maintained at zero degrees. It is the heat source $f(x, t)$ that is driving the system and causing nonzero temperatures.

The simplest way to solve (4.56)–(4.58) is to use Duhamel's principle as formulated in Section 2.5 for initial value problems on infinite domains. On bounded domains the principle is the same. It states that the solution of (4.56)–(4.58) is given by

$$u(x, t) = \int_0^t w(x, t - \tau, \tau)d\tau,$$

where $w = w(x, t, \tau)$ is the solution to the homogeneous problem

$$w_t - kw_{xx} = 0, \quad 0 < x < \pi, \ t > 0, \tag{4.59}$$

$$w(0, t, \tau) = w(\pi, t, \tau) = 0, \quad t > 0, \tag{4.60}$$

$$w(x, 0, \tau) = f(x, \tau), \quad 0 < x < \pi. \tag{4.61}$$

Recall that τ is a parameter. We have already solved this problem in Section 4.1 (see equations (4.8) and (4.9)); we have

$$w(x, t, \tau) = \sum_{n=1}^{\infty} c_n e^{-n^2 kt} \sin nx,$$

where

$$c_n = c_n(\tau) = \frac{2}{\pi} \int_0^{\pi} f(x, \tau) \sin nx dx.$$

Notice that the Fourier coefficients c_n depend on the parameter τ. So, the solution to (4.56)–(4.58) is

$$u(x, t) = \int_0^t \left(\sum_{n=1}^{\infty} c_n(\tau) e^{-n^2 k(t-\tau)} \sin nx \right) d\tau. \tag{4.62}$$

EXAMPLE

Solve (4.56)–(4.58) when $f(x, t) = \sin x$. Easily the Fourier coefficients are given by $c_n = 0, \quad n > 1, c_1 = 1$. Then the solution is

$$u(x, t) = \int_0^t c_1 e^{-k(t-\tau)} \sin x d\tau$$
$$= \frac{1}{k}(1 - e^{-kt}) \sin x.$$

In the limit as $t \to \infty$, we have $\lim_{t\to\infty} u(x, t) = (1/k) \sin x$. Notice that this is the same as the steady-state solution, i.e., the time-independent solution of the system. This is true in general: If the source term in (4.56)–(4.58) depends only on x, or $f(x, t) = F(x)$, then as $t \to \infty$ the solution to (4.56)–(4.58) approaches the steady-state solution, i.e., the solution $v = v(x)$ to

$$-kv'' = F(x), \quad v(0) = v(\pi) = 0.$$

Equations with sources can also be solved by an eigenfunction method. The first step is to find the eigenfunctions of the Sturm–Liouville problem associated with the *homogeneous* problem, with no sources. In the case of (4.56)–(4.58) the eigenvalues and eigenfunctions are (see Section 4.1)

$$\lambda_n = n^2, \quad y_n(x) = \sin nx, \quad n = 1, 2, \ldots.$$

Then we assume a solution of the inhomogeneous problem (4.56)–(4.58) of the form

$$u(x, t) = \sum_{n=1}^{\infty} g_n(t) \sin nx, \tag{4.63}$$

where the $g_n(t)$ are to be determined. We can determine the $g_n(t)$ by substituting this expression for u into the PDE (4.56), along with the expression for the eigenfunction expansion for f, namely,

$$f(x, t) = \sum_{n=1}^{\infty} f_n(t) \sin nx.$$

In this last expansion, the $f_n(t)$ are known because f is known; the $f_n(t)$ are the Fourier coefficients

$$f_n(t) = \frac{2}{\pi} \int_0^{\pi} f(x, t) \sin nx dx. \tag{4.64}$$

Carrying out this substitution, we get

$$\frac{\partial}{\partial t} \sum_{n=1}^{\infty} g_n(t) \sin nx - k \frac{\partial^2}{\partial x^2} \sum_{n=1}^{\infty} g_n(t) \sin nx = \sum_{n=1}^{\infty} f_n(t) \sin nx,$$

or

$$\sum_{n=1}^{\infty} g_n'(t) \sin nx + k \sum_{n=1}^{\infty} n^2 g_n(t) \sin nx = \sum_{n=1}^{\infty} f_n(t) \sin nx.$$

Collecting the coefficients of the independent eigenfunctions $\sin nx$, we obtain

$$g_n'(t) + n^2 k g_n(t) = f_n(t),$$

which is a linear ordinary differential equation for the unknown $g_n(t)$. It can be solved in the standard way by multiplying by the integrating factor e^{n^2kt} and then integrating both sides (see the Appendix on ODEs). After some straightforward manipulation we obtain

$$g_n(t) = g_n(0)e^{-n^2kt} + \int_0^t f_n(\tau)e^{-n^2k(t-\tau)}d\tau.$$

To obtain the values of $g_n(0)$ we substitute (4.63) into the initial condition to get

$$\sum_{n=1}^{\infty} g_n(0)\sin nx = 0,$$

which implies $g_n(0) = 0$ for all n. Therefore, we have determined the $g_n(t)$, and the solution to the problem (4.56)–(4.58) is

$$u(x,t) = \sum_{n=1}^{\infty}\left(\int_0^t f_n(\tau)e^{-n^2k(t-\tau)}d\tau\right)\sin nx, \tag{4.65}$$

where the $f_n(t)$ are given by (4.64). This solution formula (4.65) is the same as the solution formula (4.62) obtained by Duhamel's principle.

In summary, in the eigenfunction method for solving problems with sources we expand both the source f and the assumed solution u in terms of the eigenfunctions of the homogeneous problem. Substituting these expansions into the PDE leads to ODEs for the coefficients in the eigenfunction expansion for u. Solving these ODEs for the coefficients then gives the solution u as an eigenfunction expansion. Whereas Duhamel's principle is applied to initial value problems (evolution problems), the eigenfunction expansion method can be applied to all problems, including equilibrium problems associated with Laplace's equation. □

Exercises

1. Use Duhamel's principle as formulated for the wave equation in Section 2.5 to find the solution to

$$u_{tt} - c^2u_{xx} = f(x,t), \quad 0 < x < \pi, \ t > 0,$$
$$u(0,t) = u(\pi,t) = 0, \quad t > 0,$$
$$u(x,0) = u_t(x,0) = 0, \quad 0 < x < \pi.$$

Next, use the eigenfunction method to solve the problem. Hint: In the eigenfunction method you will have to solve a second-order inhomogeneous ODE;

recall from elementary differential equations that a particular solution $y_p(t)$ to the ODE

$$y'' + p(t)y' + q(t)y = f(t)$$

is given by the variation of parameters formula (see the Appendix).

2. In problem (4.56)–(4.58) assume that the heat source does not depend on time, i.e., $f(x,t) = F(x)$. By calculating the τ-integral, show how the solution (4.62) simplifies in this case. Find a formula for the solution when the heat source is $F(x) = x(\pi - x)$ and then find the limiting temperature profile as $t \to \infty$. Show that this limiting temperature profile is the same as the steady-state solution, that is, the solution $v = v(x)$ to

$$-kv''(x) = x(\pi - x), \quad 0 < x < \pi; \quad v(0) = v(\pi) = 0.$$

3. Use the eigenfunction method to solve the Dirichlet problem for the Poisson equation:

$$u_{xx} + u_{yy} = f(x,y), \quad 0 < x < \pi, \; 0 < y < 1,$$

$$u(0,y) = u(\pi,y) = 0, \quad 0 < y < 1,$$

$$u(x,0) = u(x,1) = 0, \quad 0 < x < \pi.$$

Hint: Assume $u(x,t) = \sum_{n=1}^{\infty} g_n(y)\sin nx$. Substitute into the PDE and boundary conditions at $y = 0$ and $y = 1$ to obtain

$$g_n''(y) - n^2 g_n(y) = f_n(y), \quad g_n(0) = g_n(1) = 0,$$

where $f_n(y) = \left(\frac{2}{\pi}\right)\int_0^{\pi} f(x,y)\sin nx dx$. Write the general solution in the form

$$g_n(y) = c_1 e^{ny} + c_2 e^{-ny} + \frac{2}{n}\int_0^y f_n(\xi)\sinh n(y - \xi)d\xi$$

and determine the constants c_1 and c_2.

4. Use Duhamel's principle to find the solution to

$$u_t = \Delta u + f(r,t), \quad 0 \le r < R, \; t > 0,$$

$$u(R,t) = 0, \quad t > 0,$$

$$u(r,0) = 0, \quad 0 \le r < R.$$

4.7 Parameter Identification Problems

In some physical problems all of the input parameters may not be known a priori. In a *parameter identification problem* we ask whether it is possible to take certain measurements and thereby determine an unknown parameter (a constant) or distributed parameter (a function). For example, suppose the diffusivity of a metal rod is unknown; will holding the temperature constant at one end and measuring the heat flux out the other end determine the diffusivity? Parameter identification problems are a subclass of *inverse problems*; these can be described qualitatively as problems where the input is to be recovered from the output, rather than conversely, which is the case for so-called *direct problems*. All of the problems we have encountered so far are direct problems (given all the data, find the solution), but one could argue that inverse problems play an equally important role in applied science. In many empirical problems we wish to measure the output and use that information to determine properties of the system. In this section we will not be too ambitious, but rather only introduce some of the issues. For a detailed introduction we refer the reader to the excellent monograph by Groetsch (1993).

Before studying a PDE model, we first examine a simple ODE model. Consider the total population $p = p(t)$ of animals in a fixed, closed geographic region. Suppose the initial population p_0 is known, but the growth rate r is not known. Can we determine the growth rate by measuring the population $p^* = p(t^*)$ at a later, single, instant of time t^*? Of course, we need a model. We assume the Malthus model of population growth

$$\frac{dp}{dt} = rp, \quad t > 0. \tag{4.66}$$

The initial condition is

$$p(0) = p_0, \tag{4.67}$$

and the question is, Can we recover r from the additional measurement

$$p(t^*) = p^*, \tag{4.68}$$

where t^* and p^* are known? Before solving this problem, let us make some observations. If r is known, then problem (4.66)–(4.67) is a simple, direct, well-posed initial value problem: If p_0 and r are known, then the unique solution to this direct problem can be determined exactly. Easily, $p(t) = p_0e^{rt}$, which gives exponential growth. But for the parameter identification problem, it is r that is to be found. Intuitively, it appears that an extra condition, like (4.68), is enough to determine r. Indeed, from the solution to the direct problem,

$$p^* = p(t^*) = p_0e^{rt^*}.$$

Therefore,

$$r = \frac{1}{t^*}\ln(\frac{p^*}{p_0}).$$

In all parameter identification problems we are interested in the sensitivity, or stability, of our recovery strategy. That is, are the results relatively unaffected by small errors in our measurements? Exercise 1 explores the sensitivity question in this population model.

The parameter identification problem above is simple and has a simple answer. It is not too difficult to invent a problem that is not so simple to solve, especially for PDE models.

Let us consider a long ($x \geq 0$) metal slab of unknown thermal conductivity K but known density ρ and specific heat c. For simplicity, take

$\rho = c = 1$. Suppose further that measurements can be made only at the face $x = 0$. If a known temperature $f(t)$ at $x = 0$ is applied, can we measure the heat flux at $x = 0$ at a single instant of time t_0 and thereby determine K? We assume $f(0) = 0$. The PDE model is

$$u_t = Ku_{xx}, \quad x, t > 0, \tag{4.69}$$

$$u(0, t) = f(t), \quad t > 0, \tag{4.70}$$

$$u(x, 0) = 0, \quad x > 0, \tag{4.71}$$

which is a well-posed problem. We are asking whether we can determine K from a single flux measurement

$$-Ku_x(0, t_0) = a, \tag{4.72}$$

where a is known. Using Laplace transforms we solved the direct problem (4.69)–(4.71) in Section 2.6. The solution is

$$u(x, t) = \int_0^t \frac{x}{\sqrt{4K\pi(t-\tau)^3}} e^{-x^2/(4K(t-\tau))} f(\tau)d\tau. \tag{4.73}$$

It appears that the strategy should be to calculate the flux at $(0, t_0)$ from the solution formula (4.73). Indeed this is the case, but calculating the x derivative of u is not a simple matter. The straightforward approach of pulling a partial derivative $\partial/\partial x$ under the integral sign fails because one of the resulting improper integrals cannot be evaluated at $x = 0$; it does not exist. The reader should verify this statement. Therefore, we must be more clever in calculating u_x. To this end we note that (4.73) can be written as

$$u(x, t) = -2K \int_0^t \frac{\partial}{\partial x} G(x, t-\tau) f(\tau) d\tau,$$

where $G(x, t)$ is the heat kernel

$$G(x, t) = \frac{1}{\sqrt{4\pi Kt}} e^{-x^2/(4Kt)}.$$

Therefore,

$$\begin{aligned} u_x(x, t) &= -2K \int_0^t \frac{\partial^2}{\partial x^2} G(x, t-\tau) f(\tau) d\tau \\ &= 2 \int_0^t \frac{\partial}{\partial \tau} G(x, t-\tau) f(\tau) d\tau, \end{aligned}$$

since $-G_\tau = G_t = KG_{xx}$ (the heat kernel satisfies the heat equation). Now we integrate by parts to obtain

$$u_x(x, t) = -2 \int_0^t G(x, t-\tau) f'(\tau) d\tau.$$

The boundary terms generated by the integration by parts are zero. Consequently, we have

$$-Ku_x(0, t_0) = \sqrt{K}\int_0^{t_0} \frac{f'(\tau)}{\sqrt{\pi(t_0-\tau)}}\,d\tau = a.$$

This equation uniquely determines K and solves the parameter identification problem.

For example, if $f(t) = \beta t$, i.e., the temperature is increased linearly, then the integral can be calculated exactly (Exercise 6), and we obtain

$$K = \frac{\pi a^2}{4\beta^2 t_0}. \tag{4.74}$$

The exercises contain additional examples. Often, inverse problems, and in particular distributed parameter identification problems, lead to an integral equation for the unknown function. These equations are difficult to resolve, and stability is often a problem.

Exercises

1. In this exercise we examine the sensitivity of determining the growth constant r from a measurement of the population in the example at the beginning of this section. Suppose two different measurements p_1 and p_2 of the population are made at the same time t^*. Each leads to growth constants r_1 and r_2. Show that if p_1 and p_2 are close, then r_1 and r_2 are close. In particular, show that

$$|r_1 - r_2| \le C|p_1 - p_2|,$$

where C is a constant. Thus, this recovery strategy is *stable* in the sense that a small error in measuring the population will not drastically change the growth constant.

2. This problem deals with determining the unknown density of a nonhomogeneous vibrating string by observing one fundamental frequency and the corresponding mode of vibration. Consider a stretched string of unit length and unit tension that is fastened at both ends. The displacement is governed by (see Section 1.5)

$$\rho(x)u_{tt} = u_{xx}, \quad 0 < x < 1,\ t > 0,$$

$$u(0, t) = u(1, t) = 0, \quad t > 0,$$

where $\rho(x)$ is the unknown density. If we assume a solution of the form $u(x, t) = y(x)g(t)$, then we obtain a Sturm–Liouville problem

$$-y'' = \lambda\rho(x)y, \quad 0 < x < 1; \quad y(0) = y(1) = 0.$$

Suppose we observe (say, using a strobe light) a fundamental frequency $\lambda = \lambda_f$ and the associated mode (eigenfunction) $y = y_f(x)$. Show that the unknown density must satisfy the integral equation

$$\int_0^1 (1-x)y_f(x)\rho(x)dx = \frac{y_f'(0)}{\lambda_f}.$$

(Hint: Integrate the ODE from $x = 0$ to $x = s$ and then from $s = 0$ to $s = 1$). Determine the density if it is a constant.

3. Consider heat flow in a rod of length π and unit diffusivity whose ends are held at constant zero temperature and whose initial temperature is zero degrees. Suppose further that there is an external heat source $f(x)$ supplying heat to the bar. Address the question of recovering the heat source $f(x)$ from a temporal measurement of the temperature $U(t)$ made at the midpoint of the rod. Proceed by formulating the appropriate equations and show that $f(x)$ must be the solution to the integral equation

$$U(t) = \int_0^t \int_0^\pi g(\xi, t-\tau)f(\xi)\, d\xi\, d\tau,$$

where

$$g(\xi, t) = \frac{2}{\pi}\sum_{n=1}^{\infty} e^{-n^2 t} \sin\frac{n\pi}{2} \sin n\xi.$$

Show that for any positive integer m, the pair

$$u(x,t) = m^{-3/2}(1-e^{-m^2 t})\sin mx, \quad f(x) = \sqrt{m}\sin mx$$

satisfies the model equations you proposed. Show that for m large, $u(x, t)$ is (uniformly) small, yet $\max|f(x)|$ is large, and thus state why small errors in the measurement $U(t)$ may lead to large differences in $f(x)$. (This means that recovery of f by solving the integral equation above may be unstable and therefore difficult.)

4. Suppose a chemical reactor occupying the space $x < 0$ operates at some unknown temperature $F(t)$. To determine $F(t)$ we insert a long, laterally insulated metal probe of unit diffusivity, and we measure the temperature $U(t)$ of the probe at $x = 1$. Assume that the probe occupies $0 \le x < \infty$. Formulate the model equations and show that $F(t)$ satisfies the integral equation

$$U(t) = \frac{1}{2\sqrt{\pi}}\int_0^t \frac{F(\tau)}{(t-\tau)^{3/2}} e^{-1/4(t-\tau)} d\tau.$$

Hint: Solve the direct problem by Laplace transforms and use the table entry

$$\mathcal{L}^{-1}(e^{-\sqrt{s}x}) = \frac{x}{2\sqrt{\pi}t^{3/2}} e^{-x^2/(4t)}.$$

Suppose the temperature of the reactor $F(t)$ is a constant F_0; find F_0 if the temperature at $x = 1$ is found to be 10 degrees at $t = 5$.

5. It is believed that nerve impulses are transmitted along axons by both diffusion and convection. In fundamental experiments the biologists Hodgkin and Keyes used radioactive potassium ^{42}K to measure how ions convect and diffuse in squid axons. If $u = u(x, t)$ is the concentration of potassium in a long axon, then the convection–diffusion model is

$$u_t = Du_{xx} - vu_x, \quad x \in \mathbf{R}, \quad t > 0.$$

The velocity v can be measured directly, but the diffusion constant D is difficult to measure. To determine D take a known initial concentration $u(x, 0) = e^{-x^2/a}$ and solve for the concentration $u(x, t)$. Obtain

$$u(x, t) = \frac{\sqrt{a}}{\sqrt{a + 4Dt}} e^{-(x-vt)^2/(a+4Dt)}.$$

Show how D can be recovered from a continuous measurement $U(t) = u(x_0, t)$ of the potassium concentration at some fixed location $x = x_0$.

6. Verify (4.74).

4.8 Finite Difference Methods

Up until now we have studied PDEs from an analytical viewpoint, and we have observed how a computer algebra package like Maple or Mathematica can aid in these studies. Now we devote a brief section to solving PDEs numerically. In most cases (for example, in industry and practical science), PDEs are almost always solved numerically on a computer. This is because most real-world problems are too complicated to solve analytically. And even if a problem can be solved analytically, usually the solution is in the form of a difficult integral or an infinite series, thereby forcing a computer calculation anyway. Why not just begin with a computational approach in the first place? The reader may wonder why we have devoted so much space to analytical calculations and so little to computational issues. The reason for our approach is that this is a text about understanding the concepts that underpin the subject; it is just as important to develop one's insights and understanding as it is to compute a solution. A numerical analyst would not attempt to solve PDEs on a computer without some basic understanding about their nature and about what to expect from solutions. With the analytical tools we have developed, the reader will have the foundation to study numerical algorithms. A good, brief introduction to finite difference methods can be found in Morton and Mayers (1994).

In this section we introduce a basic numerical procedure to solve PDEs. By a numerical or computational method we mean something different from asking a computer algebra program to return a formula. By a numerical method we generally mean that the continuous PDE problem is replaced by a discrete problem that can be solved on a computer in finitely many steps. The result is a discrete solution, where the solution surface is known approximately at only finitely many points. In this section we focus attention on the *finite difference method* for the diffusion equation and for Laplace's equation. The idea is to replace the partial derivatives in the equation by their difference-quotient approximations and then let the computer solve the resulting difference equation. The reader should be familiar with this strategy for ordinary differential equations. For example, to numerically solve the initial value problem

$$y' = f(t, y), \quad 0 < t < 1; \quad y(0) = y_0,$$

we can discretize the interval $0 \leq t \leq 1$ by defining a finite number of discrete points $t_n = nh$, $n = 0, 1, \ldots, N$, where h is the *step size*. Here $h = 1/N$. At each point t_n we seek to determine a value Y_n that is a good approximation to the exact value $y(t_n)$ of the solution. From differential calculus we know that $y'(t_n)$ can be approximated by a forward difference quotient

$$y'(t_n) \approx \frac{y(t_{n+1}) - y(t_n)}{h},$$

provided that h is small. Consequently, we approximate the differential equation by

$$\frac{Y_{n+1} - Y_n}{h} = f(t_n, Y_n), \quad n = 0, 1, \ldots, N,$$

or

$$Y_{n+1} = Y_n + hf(t_n, Y_n), \quad n = 0, 1, \ldots, N.$$

The reader may recognize this difference equation as defining *Euler's method*; given $Y_0 = y_0$, it provides a numerical algorithm to calculate the approximations $Y_1, Y_2, \ldots$ recursively. This same strategy, the idea of marching forward in time, goes over to evolution problems in PDEs.

We will illustrate the finite difference method for an initial boundary value problem for the diffusion equation

$$u_t = Du_{xx}, \quad 0 < x < 1, \quad t > 0, \tag{4.75}$$

$$u(0, t) = u(1, t) = 0, \quad t > 0, \tag{4.76}$$

$$u(x, 0) = f(x), \quad 0 < x < 1. \tag{4.77}$$

The first step to discretize the region of space–time where we want to obtain a solution. In this case the region is $0 \leq x \leq 1$, $0 \leq t \leq T$. We have put a bound on time because in practice we only solve a problem up until a finite time. Discretizing means defining a lattice of points in this space–time region by

$$x_j = jh, \quad t_n = nk, \quad j = 0, 1, \ldots, J; \quad n = 0, 1, \ldots, N,$$

where the fixed numbers h and k are the spatial and temporal step sizes, respectively. Here, $h = 1/J$ and $k = 1/N$. The integer J is the number of subintervals in $0 \leq x \leq 1$, and N is the number of time steps to be taken. Figure 4.7 shows the lattice of points (the lattice points are also called grid points or nodes). At each node (x_j, t_n) of the lattice we seek an approximation, which we call U_j^n, to the exact value $u(x_j, t_n)$ of the solution. Note that the superscript n refers to time, and the subscript j refers to space. We regard U_j^n as a two-dimensional array, or matrix, where n is a row index and j is a column index. To obtain equations for the U_j^n, we replace the partial derivatives in the PDE by their difference approximations. From calculus we know

$$u_t(x_j, t_n) \approx \frac{u(x_j, t_{n+1}) - u(x_j, t_n)}{k}$$

and

$$u_{xx}(x_j, t_n) \approx \frac{u(x_{j-1}, t_n) - 2u(x_j, t_n) + u(x_{j+1}, t_n)}{h^2}.$$

The first of these formulas is the usual forward difference approximation for a first derivative, and the second is a difference approximation for a second derivative; the latter follows from a Taylor approximation exactly as derived in Section 1.8. So the PDE (4.75) at the point (x_j, t_n) is replaced by the difference equation

$$\frac{U_j^{n+1} - U_j^n}{k} = D\frac{U_{j-1}^n - 2U_j^n + U_{j+1}^n}{h^2},$$

or, upon solving for U_j^{n+1},

$$U_j^{n+1} = U_j^n + \frac{kD}{h^2}(U_{j-1}^n - 2U_j^n + U_{j+1}^n). \tag{4.78}$$

Observe that this equation relates the approximate values of the solution at the four points $(x_{j-1}, t_n), (x_j, t_n), (x_{j-1}, t_n), (x_j, t_{n+1})$. These four points form the *computational atom* for the difference scheme (see Figure 4.7).

The difference equation (4.78) gives the approximate solution at the node (x_j, t_{n+1}) in terms of approximations at three earlier nodes. Now we see how to fill up the lattice with approximate values. We know the values at $t = 0$ from the initial condition (4.77). That is, we know

$$U_j^0 = f(x_j), \quad j = 0, 1, \ldots, J.$$

¿From the boundary conditions (4.76) we also know

$$U_0^n = 0, \quad U_J^n = 0, \quad n = 1, 2, \ldots, N.$$

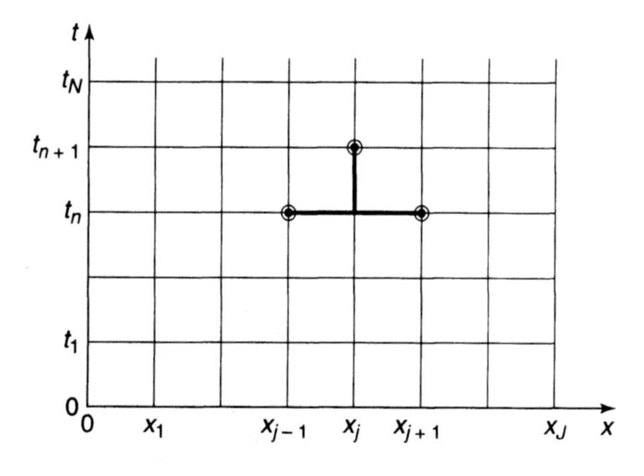

Figure 4.7. The discrete lattice and the computational atom for the diffusion equation.

The difference formula (4.78) can now be applied at all the interior lattice points, beginning with the values at the $t = 0$ level, to compute the values at the $t = t_1$ level, then using those to compute the values at the $t = t_2$ level, and so on. Therefore, using the difference equation, we can march forward in time, continually updating the previous temperature profile. Think of filling out the array U_j^n, row by row.

A finite difference scheme, or algorithm, like (4.78) is called an *explicit scheme* because it permits the explicit calculation of an approximation at the next time step in terms of values at a previous time step. Because of the error (called the truncation error) that is present in (4.78) due to replacing derivatives by differences, the scheme is not always accurate. If the time step k is too large, then highly inaccurate results will be obtained. It can be shown that we must have the *stability condition*

$$\frac{kD}{h^2} \leq \frac{1}{2}$$

for the scheme to work; this condition limits the time step. Some experiments to illustrate this behavior are suggested in the exercises.

EXAMPLE

It is easy to program the explicit algorithm (4.78) described above to calculate approximate temperatures in a rod. The student is invited to write a program in Matlab, FORTRAN, or C++ to carry out this computation. Let us describe a simple Maple program to do the same. Let the initial temperature be given by $f(x) = x^3(1 - x)$. We store the values of U_j^n in a two-dimensional array so that we can use `matrixplot` to graph the solution surface. The program is listed and the solution surface is plotted in Figure 4.8. □

This same type of marching procedure works for initial value problems associated with the wave equation:

$$u_{tt} = c^2 u_{xx}, \quad 0 < x < 1, \ t > 0, \tag{4.79}$$

$$u(0, t) = u(1, t) = 0, \quad t > 0, \tag{4.80}$$

$$u(x, 0) = f(x), \quad u_t(x, 0) = g(x), \quad 0 \leq x \leq 1. \tag{4.81}$$

Using the same lattice as for the diffusion problem, we can approximate the PDE by

$$\frac{U_j^{n-1} - 2U_j^n + U_j^{n+1}}{k^2} = c^2 \frac{U_{j-1}^n - 2U_j^n + U_{j+1}^n}{h^2},$$

which gives

$$U_j^{n+1} = 2U_j^n - U_j^{n-1} + \left(\frac{ck}{h}\right)^2 (U_{j-1}^n - 2U_j^n + U_{j+1}^n).$$

The computational atom is shown in Figure 4.9. To compute the value at the $(n + 1)$st time step we now need values from the previous *two* time

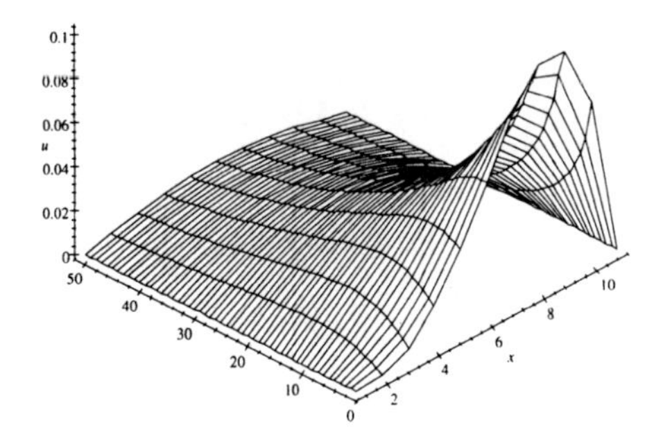

Figure 4.8a. A graph of the temperature surface when $f(x) = x^3(1-x)$.

Set the discretization:

```
J:=10: h:=1/J: k:=0.004: N:=50: r:=k/h^2:
```

Define the array and set the initial condition:

```
U:=array(1..J+1,1..N+1):
f:=x- >x^3*(1-x):
for j from 1 to J+1 do U[j,1]:=f((j-1))*h): od:
```

Set the Dirichlet boundary conditions:

```
for n from 1 to N+1 do
U[1,n]:=0.0: U[J+1,n]:=0.0: od:
```

Fill up the array using the difference equation:

```
for n from 1 to N do:
for j from 2 to J do:
U[j,n+1]:=U[j,n]+r*(U[j-1,n]-2*U[j,n]+U[j+1,n]):
od: od:
```

Plot the solution array:

```
with(linalg): with(plots):
matrixplot(U,labels=['x','t','u']);
```

Figure 4.8b. Maple program listing to solve (4.75)–(4.77) when $f(x)=x^3(1-x)$, $D=1$.

steps. Therefore, to start the marching scheme we require values from the first two ($t_0 = 0$ and $t_1 = k$) time rows. The $t = 0$ row is given by the initial condition $u(x, 0) = f(x)$; so we have

$$U_j^0 = f(x_j), \quad j = 0, 1, \ldots, J.$$

The $t = t_1$ time row can be computed using the initial velocity condition $u_t(x, 0) = g(x)$, which we approximate by

$$\frac{U_j^1 - U_j^0}{k} = g(x_j), \quad j = 1, 2, \ldots, J-1.$$

Now we have all the ingredients to start the scheme and march forward in time to fill up the lattice with values. Again there is a stability condition, namely,

$$c \le \frac{h}{k}.$$

This is called the Courant–Friedrichs–Lax (CFL) condition. Physically, it states that the speed h/k of which lattice points are being calculated must exceed the speed c of which waves are propagated in the system. Thus the calculated lattice points will contain all of the information about the wave.

In the exercises we ask the reader to compute the solution to a wave-like problem.

Now we solve a BVP for Laplace's equation in the unit square with given values on the boundary of the square. This Dirichlet problem is a pure boundary value problem, and we should not expect a marching-type scheme to work; there is no guarantee that when we march from one boundary in the problem to another we will reach the specified values at that boundary. This means that we must solve for the approximate values at all of the lattice points simultaneously. To illustrate the procedure we consider the problem

$$u_{xx} + u_{yy} = 0, \quad 0 < x < 1, 0 < y < 1, \tag{4.82}$$

$$u(0, y) = u(1, y) = 0, \quad 0 < y < 1, \tag{4.83}$$

$$u(x, 0) = f(x), \quad u(x, 1) = 0, \quad 0 \le x \le 1. \tag{4.84}$$

We discretize the unit square by defining lattice points

$$x_i = ih, \quad y_j = jk, \quad i = 0, 1, \dots, I; \quad j = 0, 1, \dots, J,$$

where $h = 1/I$ and $k = 1/J$ are the fixed step sizes in the x and y directions, respectively. We let $U_{i,j}$, with two subscripts, denote the approximate value of $u(x_i, y_j)$. Then Laplace's equation can be approximated by the difference equation

$$\frac{U_{i-1,j} - 2U_{i,j} + U_{i+1,j}}{h^2} + \frac{U_{i,j-1} - 2U_{i,j} + U_{i,j+1}}{k^2} = 0.$$

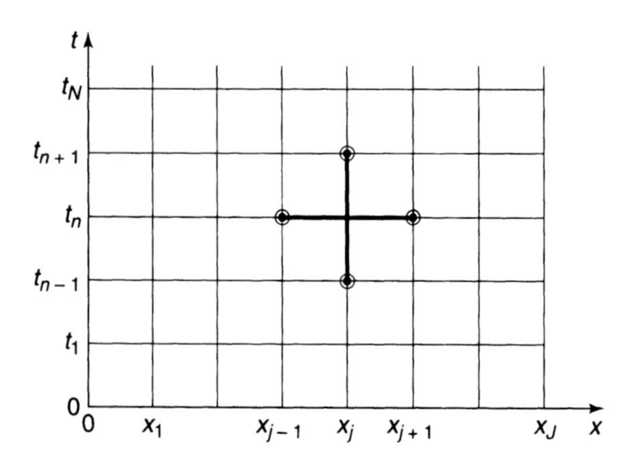

Figure 4.9. Computational atom for the wave equation.

The reader should review Section 1.8, where a similar approximation is made. The difference equation can be rewritten as

$$U_{i,j} = \frac{k^2}{2k^2 + 2h^2}(U_{i-1,j} + U_{i+1,j}) + \frac{h^2}{2k^2 + 2h^2}(U_{i,j-1} + U_{i,j+1}). \quad (4.85)$$

The computational atom is shown in Figure 4.10. The difference equation relates the node at the center of the atom to four adjacent nodes. As we observed in Section 1.8, if $h = k$, then this difference equation states that the value at the center node is approximately the average of the values at the four adjacent nodes; if $h \neq k$, the approximate value at the center is a weighted average of the adjacent values.

Now the strategy is this. The values of $U_{i,j}$ are known at the nodes on the boundary of the square. We want to find the values at the $(I - 1) \times (J - 1)$ interior lattice points. Therefore, we apply the difference equation (4.85) to each interior lattice point, i.e., for $i = 1, \ldots, I-1;\ j = 1, \ldots, J-1$. The result is a linear-algebraic system consisting of $(I - 1) \times (J - 1)$ equations in the $(I - 1) \times (J - 1)$ unknowns $U_{i,j}$. Therefore, we have reduced the Dirichlet problem to a linear algebra problem.

Example

In (4.82)–(4.84) take $f(x) = x^3(1 - x)$. To illustrate the procedure take $I = 4$ and $J = 3$; thus $j = 1/4$ and $k = 1/3$. See Figure 4.11. Then

$$\frac{h^2}{2k^2 + 2h^2} = 0.18, \qquad \frac{k^2}{2k^2 + 2h^2} = 0.32.$$

There are six interior lattice points where the solution is to be found. By the boundary conditions, the values along the top and two sides are zero, and the values along the lower boundary are $U_{0,0} = 0$, $U_{1,0} = 0.0117$, $U_{2,0} = 0.0625$, $U_{3,0} = 0.1054$, $U_{4,0} = 0$. Now we apply the difference equation (4.85) successively at the six points $(i, j) =$

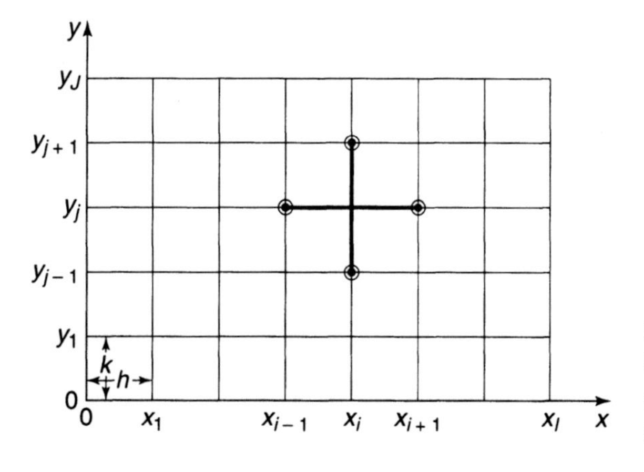

Figure 4.10. Computational atom for Laplace's equation.

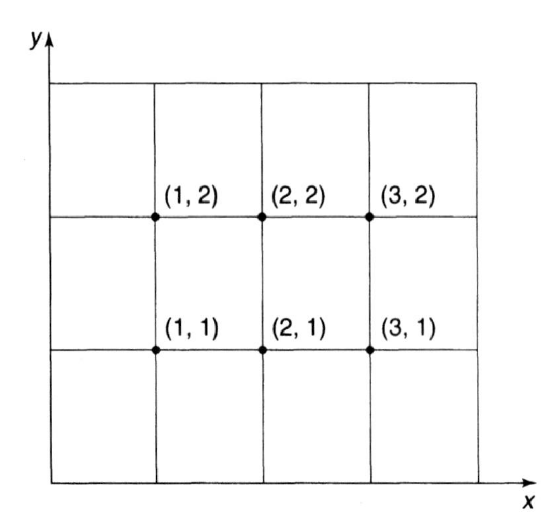

Figure 4.11. Discrete lattice.

(1, 1), (1, 2), (2, 1), (2, 2), (3, 1), (3, 2) to obtain the six equations

$$U_{1,1} = 0.32(0 + U_{2,1}) + 0.18(0.0117 + U_{1,2}),$$

$$U_{1,2} = 0.32(0 + U_{2,2}) + 0.18(U_{1,1} + 0),$$

$$U_{2,1} = 0.32(U_{1,1} + U_{3,1}) + 0.18(0.0625 + U_{2,2}),$$

$$U_{2,2} = 0.32(U_{1,2} + U_{3,2}) + 0.18(U_{2,1} + 0),$$

$$U_{3,1} = 0.32(U_{2,1} + 0) + 0.18(0.1054 + U_{3,2}),$$

$$U_{3,2} = 0.32(U_{2,2} + 0) + 0.18(U_{3,1} + 0).$$

We can solve (we used a TI-86 calculator) these six equations to obtain the approximate solution at the six interior nodal points:

$$U_{1,1} = 0.007, U_{1,2} = 0.003, U_{2,1} = 0.015,$$

$$U_{2,2} = 0.006, U_{3,1} = 0.024, U_{3,2} = 0.006.$$

□

One way to solve a linear system is by an iterative process known as the *Gauss–Seidel method*. In this method we initialize the array $U_{i,j}$ of unknowns and then systematically cycle through all the interior lattice points, replacing the old estimates with new ones calculated by the difference equation (4.85). After a number of cycles, the estimates will converge to the solution of the linear system. One could terminate the iterations when the change, measured in some manner, from the previous values is small. This process is extremely simple to program. The cycling can be accomplished in a few lines of code. For example, after fixing the boundary values, we can cycle through the interior points N times using the following Maple commands:

```
for n from 1 to N do:
```

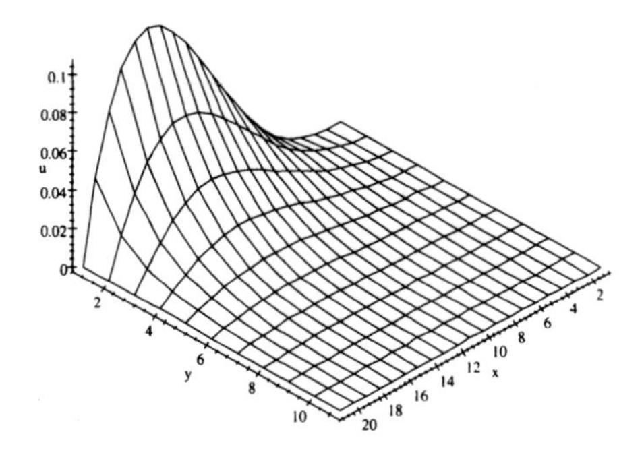

Figure 4.12a. A graph of the solution to the Dirichlet problem (4.82)-(4.84) when $f(x) = x^3(1-x)$.

Set the discretization:

```
II:=20: J:=10: N:=20: h:=1/II: k:=1/J:
r:=0.5*k^2/(k^2+h^2): s:=0.5*h^2/(k^2+h^2):
U:=array(1..II+1,1..J+1):
```

Set the boundary conditions:

```
f:=x- >x^3*(1-x):
for i from 1 to II+1 do U[i,1]:=f((i-1)*h): od:
for i from 1 to II+1 do U[i,J+1]:=0.0: od:
for j from 2 to J do U[1,j]:=0.0: od:
for j from 2 to J do U[II+1,j]:=0.0: od
```

Initialize the array at the interior points:

```
for i from 2 to II do:
for j from 2 to J do:
U[i,j]:=0.0: od: od:
```

Carry out the iteration procedure:

```
for n from 1 to N do:
  for i from 2 to J do:
  for j from 2 to J do:
  U[i,j]:=r*(U[i-1,j]+U[i+1,j])
    +s*(U(i,j-1]+U[i,j+1]):
  od: od:
od:
```

Plot the solution array:

```
with(linalg): with(plots):
matrixplot(U,labels=['x','y','u']);
```

Figure 4.12b. Maple program listing to solve the Dirichlet problem (4.82)-(4.84) when $f(x) = x^3(1-x)$.

```
for i from 1 to I-1 do:
for j from 1 to J-1 do:
U[i,j]:=k^2/(2*k^2+2*h^2)*(U[i-1,j]+U[i+1,j])
+h^2/(2*k^2+2*h^2)*(U[i,j-1]+U[i,j+1]):
od: od: od:
```

Figure 4.12 lists a Maple program to solve the Dirichlet problem in the last example using $h = 0.05$ and $k = 0.1$. Thus $I = 20$ and $J = 10$, giving $19 \times 9 = 171$ interior lattice points. Figure 4.12 also shows the solution surface.

This section was only a brief introduction to one method, the finite difference method. And we have not always presented the best method. Numerical methods for PDEs is one of the most active research areas in applied mathematics, computer science, and the applied sciences as investigators seek to find faster, more accurate algorithms. The reader will find a large amount of literature on the subject.

Exercises

1. Use Euler's method with step size $h = 0.1$ to numerically solve the initial value problem $y' = -2ty + y^2, \quad y(0) = 1$ on the interval $0 \le t \le 2$. Compare your approximations with the exact solution.

2. Run the Maple program listed in Figure 4.8 with time steps of $k = 0.1$ and $k = 0.01$. Is the stability condition $k/h^2 \le 0.5$ satisfied?

3. Develop an explicit scheme to solve the heat conduction problem

$$u_t = u_{xx}, \quad 0 < x < 1, \quad t > 0,$$

$$u(0,t) = 1, \quad u_x(1,t) = -(u(1,t) - g(t)), \quad t > 0,$$

$$u(x,0) = 0, \quad 0 < x < 1.$$

Hint: Approximate the radiation condition $u_x(1,t) = -(u(1,t) - g(t))$ by

$$\frac{u(x_J, t_n) - u(x_{J-1}, t_n)}{h} = -(u(x_J, t_n) - g(t_n)).$$

Pick $g(t) = 1$ and compute an approximate solution surface.

4. Write a program in some programming language or in a computer algebra package to solve the hyperbolic problem (4.82)–(4.84) when $f(x) = 0$ and $g(x) = x(1 - x)$ with $c = 0.25$. Take $h = 0.125$ and experiment with various time step sizes k and illustrate the validity of the CFL condition. Compare with the exact solution.

5. Write a finite difference program to solve the following steady-state problems:

$$u_{xx} + u_{yy} = 0, \quad 0 < x < 2,\ 0 < y < 1,$$
$$u(0, y) = 2 \sin \pi y, \quad u(2, y) = 0, \quad 0 < y < 1,$$
$$u(x, 0) = x(2 - x), \quad u(x, 1) = 0, \quad 0 < x < 2.$$

$$u_{xx} + u_{yy} = 0, \quad 0 < x < 2,\ 0 < y < 1,$$
$$u_x(0, y) = 0, \quad u(2, y) = 0, \quad 0 < y < 1,$$
$$u(x, 0) = x(2 - x)^2, \quad u(x, 1) = 0, \quad 0 < x < 2.$$

To handle the insulated boundary condition at $x = 0$ in the second problem, use a forward difference approximation; note that you must calculate the solution along that boundary.

Partial Differential Equations in the Life Sciences

5.1 Age-Structured Models

In Section 1.4 we introduced simple advection and diffusion models to describe the motion of organisms, cells, and chemicals in a biological science context. In this chapter we extend these ideas to more complicated phenomena involving age structure of a population, the propagation of epidemic waves, and the relationship between spatial pattern formation and chemical instability. These advanced models will show why PDEs have vast application in the life sciences. The mathematical methods we introduce to analyze these problems will extend the ideas and techniques presented in the earlier chapters. The reader can find extensive applications of PDEs to life science problems in Edelstein-Keshet (1988), Kot (2001), Britton (2003), and Murray (2003).

By an *age-structured model* we mean a demographic model where the population at time t has an age distribution superimposed on it. Thus, at time t we also consider the ages of the individuals. Age-structured models are akin to more general *physiologically structured models* where any other variable, say size or weight, could replace age. For example, in many non-mammalian populations the evolution of the population, especially the mortality rate, certainly depends upon the size of the animals; the survival probability is small for young fish or insects. Our modeling will focus on age structure, but similar ideas apply to any physiological structure.

We begin by considering a population of female organisms whose age-structure at time $t = 0$ is given by $f(a)$. That is, $f(a)da$ is approximately the number of females between age a and $a+da$. Generally, we use females in demographic models because they have a well-defined beginning and end to their reproductive capacity. And, for simplicity, we write the age range as $0 \le a < \infty$ even though the age of death is finite. Given the initial age structure, the mortality rate, and the fecundity rate (average offspring produced per female), the problem is to determine the age structure $u = u(a, t)$ at any time $t > 0$. Here, $u(a, t)da$ represents the number of females at time t between the ages a and $a + da$. Observe that the total female population at time t is

$$N(t) = \int_0^\infty u(a, t)da.$$

What makes this problem more intractable than other problems is that $u(0, t)$, which is the number of newborns at time t, is not known *a priori*. Rather, $u(0, t)$ depends upon the population and age of reproducing females, which is the unknown in the problem. Specifically, we assume that the *per capita mortality rate* is given as a function of age as $m = m(a)$. And we assume that the fecundity rate is $b(a, t)$, which depends upon both time and age of the female. Precisely, $b(a, t)$ is the average number of offspring per female of age a at time t; the function $b(a, t)$ is called the *maternity function*. We expect $b(a, t)$ to be zero until the age of onset of maturity and zero after menopause. In between these ages the fecundity rate varies according to age; for example, in humans, women of age 25 are more fertile than women of age 40. Figure 5.1 depicts the evolution of age structure profiles in a three-dimensional *atu*-space (age–time–density space).

The governing dynamics is given by the equation

$$u_t = -u_a - m(a)u.$$

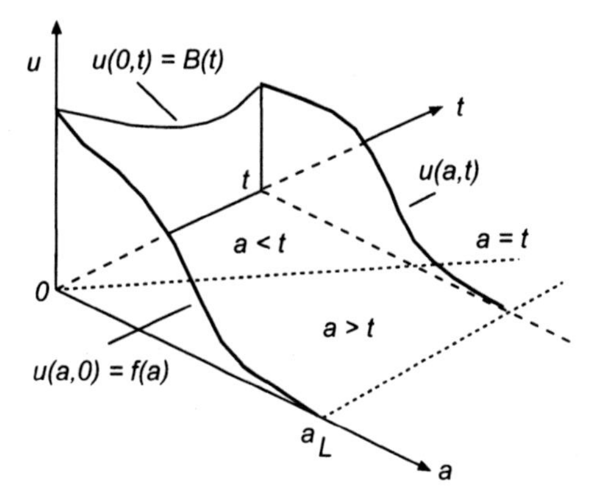

Figure 5.1. Age-structured model. Here, $f(a)$ is the initial, known age structure, and $u(0, t) = B(t)$ is the unknown offspring at age $a = 0$ and time t. The age structure $u(a, t)$ for $a > t$ is affected only by the initial population $f(a)$, whereas for $a < t$ it is affected by the entire population and its fecundity. a_L is the maximum lifetime age and individuals follow paths, or characteristics, $a = t+$ constant in age–time space.

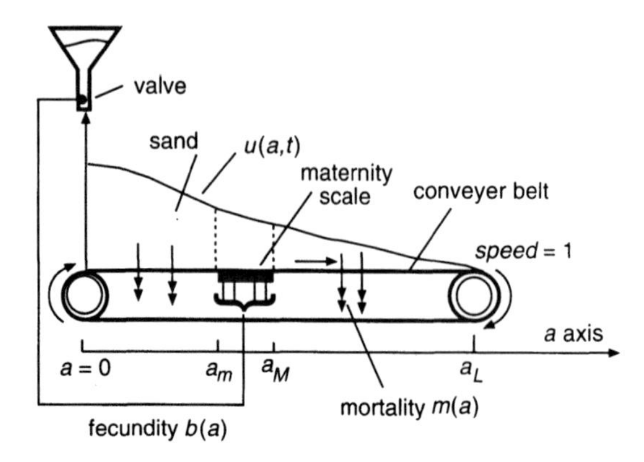

Figure 5.2. Conveyer belt visualization of an evolving age structure.

Thus, the model equation is the advection equation with speed one and sink term given by the mortality rate; note that the flux is $\phi = u$, or the number crossing the age a line at time t is just $u(a, t)$, the density at that age. This continuous time model has its origins in the work of McKendrick in 1926 and in subsequent studies by von Foerster in 1959. The equation is often called the *McKendrick–von Foerster equation*. Now we formulate the boundary condition $u(0, t)$, the number of newborns at time t. Since $u(a, t)da$ is the number of females between a and $a + da$, and b is the average reproduction rate, we have

$$\text{Number of offspring produced by females in age } a \text{ to } a + da = b(a, t)u(a, t)da$$

Thus, the total number of offspring produced by all females is the (sum) integral over all ages, or

$$u(0, t) = \int_0^\infty b(a, t)u(a, t)da.$$

Therefore we have derived the age-structured model

$$u_t = -u_a - m(a)u, \quad a > 0, \quad t > 0, \tag{5.1}$$

$$u(0, t) = \int_0^\infty b(a, t)u(a, t)da, \quad t > 0, \tag{5.2}$$

$$u(a, 0) = f(a), \quad a \geq 0. \tag{5.3}$$

We emphasize again that the left boundary condition at age $a = 0$ is not known, but rather depends upon the solution $u(a, t)$, which is also unknown. This type of condition is called a *nonlocal boundary condition* because it depends upon the integrated unknown solution in the problem.

Age-structured models can be represented visually as a conveyor belt moving at speed one (figure 5.2). Grains of sand on the belt represent the female population density, and the belt contains holes through which the sand falls, representing mortality. A scale is located between the ages of fertility, $a_m < a < a_M$; it weighs the female population and sends an signal to the valve in the funnel at $a = 0$, which releases newborns to the population at $a = 0$. At $a = a_L$, the maximum lifetime age and length of the belt, the population density is zero.

Stable age structure

Rather than at first attempting to solve (5.1)–(5.3) directly, we first take a different tack that is common in demographic models. We ignore the initial condition (5.3) and ask what happens over a long time. Births from the initial population $f(a)$ only affect the solution for a finite time because those individuals and their offspring die. Therefore, in the case that the maternity function is independent of time, i.e., $b = b(a)$, we look for a *stable age structure* of the form

$$u(a,t) = U(a)e^{rt}, \quad t \text{ large}$$

where $U(a)$ is an unknown age structure and r is an unknown growth rate. Substituting into the PDE (5.1) and making reductions gives an ODE for $U(a)$,

$$U'(a) = -(m(a) + r)U(a).$$

This equation can be solved by separation of variables to get

$$U(a) = Ce^{-ra}e^{-\int_0^a m(s)ds},$$

where C is a constant. Letting $S(a) = \exp\left(-\int_0^a m(s)ds\right)$ denote the *survivorship function*, which is the probability of surviving to age a, we can write the long time solution as

$$u(a,t) = Ce^{rt-ra}S(a). \tag{5.4}$$

To determine the growth constant r, we substitute (5.4) into the nonlocal boundary condition (5.2) to obtain

$$1 = \int_0^\infty b(a)e^{-ra}S(a)da, \tag{5.5}$$

which is the *Euler–Lotka equation*. Using numerical methods this equation can be solved for r, and we will have determined, up to a constant, the long-time age structure $U(a)$ of the population and its growth rate r.

In the special case m = constant, the Euler–Lotka equation is

$$1 = \int_0^\infty b(a)e^{-(r+m)a}da.$$

The exercises will request calculations for specific population data.

The renewal equation

The method of characteristics introduced in Section 1.2 can be used to study (5.1)–(5.3) in the simple case when $b = b(a)$ and m = constant. The PDE (5.1) is

$$u_t = -u_a - mu, \quad a > 0, \quad t > 0. \tag{5.6}$$

If we change independent variables via the formulae

$$\xi = a - t, \quad \tau = t,$$

then the PDE (5.6) becomes

$$U_\tau = -mU, \quad \text{where} \quad U = U(\xi, t).$$

This equation has general solution

$$U(\xi, \tau) = C(\xi)e^{-m\tau},$$

where C is an arbitrary function. In terms of the original variables,

$$u(a, t) = C(a - t)e^{-mt}.$$

To determine the arbitrary function C we consider two cases, $a > t$ and $a < t$. See figure 5.1. The arbitrary function will turn out different in each case. The solution in $a > t$ is determined by the initial age structure and we have

$$u(a, 0) = C(a) = f(a).$$

Therefore

$$u(a, t) = f(a - t)e^{-mt}, \quad a > t. \tag{5.7}$$

For $a < t$ let us denote, for simplicity, $B(t) = u(0, t)$. Then, applying the boundary condition gives

$$u(0, t) = B(t) = C(-t)e^{-mt},$$

or

$$C(s) = B(-s)e^{-ms}.$$

Consequently,

$$u(a, t) = B(t - a)e^{-ma}, \quad a < t. \tag{5.8}$$

The solution to (5.1)–(5.3) in the case $m(a) = m$ and $b(a, t) = b(a)$ is given by (5.7)–(5.8), but B is still unknown. To find B we substitute the expressions (5.7)–(5.8) into the yet unused nonlocal boundary condition (5.2), after breaking up the integral into two. We obtain

$$\begin{aligned} B(t) &= \int_0^\infty b(a)u(a, t)da \\ &= \int_0^t b(a)u(a, t)da + \int_t^\infty b(a)u(a, t)da, \end{aligned}$$

or

$$B(t) = \int_0^t b(a)B(t - a)e^{-ma}da + \int_t^\infty b(a)f(a - t)e^{-mt}da. \tag{5.9}$$

Equation (5.9) is a *linear integral equation* for the unknown $B(t)$, and it is called the *renewal equation*. Once it is solved for $B(t)$, then (5.7)–(5.8) give the age structure for the population. In Exercise 2 the reader is asked to solve the renewal equation in a special case. Generally, (5.9), a nonhomogeneous Volterra equation, is difficult to solve and must be dealt with numerically, or by successive approximation (iteration).

Structured predator–prey model

The previous example illustrates how age structure leads to an advection equation that can be solved, in theory, by standard change of variable methods. Next we consider a more difficult nonlinear predator–prey problem where a different technique is illustrated. The method is called the *method of moments*, which is akin to an energy method. We may add this important technique to our analytic tool bag for dealing with PDEs.

We consider a population of prey with age density $u(a,t)$ and constant *per capita* mortality rate m. Then, as above, the governing age–time dynamics is given by

$$u_t = -u_a - mu, \quad a > 0, \quad t > 0, \tag{5.10}$$

where the initial number of prey is

$$u(a,0) = f(a). \tag{5.11}$$

The total prey population is

$$N(t) = \int_0^\infty u(a,t)da.$$

We assume that the maternity function is

$$b(a) = b_0 a e^{-\gamma a}.$$

Then the prey produce offspring (eggs) given by

$$B(t) = \int_0^\infty b_0 a e^{-\gamma a} u(a,t)da.$$

Now let us introduce a total predator population $P = P(t)$; we shall not consider age-structure in this population, but only that the predators eat the eggs of the prey. (To stimulate thinking about this model, remember one of the theories posed for the extinction of the dinosaurs is that egg-eating predators caused the demise.) Because predators eat only eggs ($a = 0$), the PDE is unaffected. What is affected is the actual number of offspring $u(0,t)$ produced. Thus we no longer have $u(0,t) = B(t)$, but rather we must include a predation term that decreases the egg population. The simplest model is the Lotka–Volterra model, which requires that the number eggs eaten is proportional to the product of the number of eggs and the number of predators. That is, we have

$$u(0,t) = B(t) - kB(t)P(t),$$

where k is a proportionality constant. But the right side can be negative, so we define $M(B,P) = \max(B - kBP, 0)$ and take the number of eggs at

$a = 0$ to be

$$u(0, t) = M(B, P). \tag{5.12}$$

This equation provides the boundary condition to the problem. Finally, we ask how the predator population changes. We assume Lotka–Volterra dynamics

$$\frac{dP}{dt} = -\delta P + cBP, \tag{5.13}$$

where δ is the per capita mortality rate and c is the yield. Hence, in the absence of eggs, predators die out. Initially, we take $P(0) = P_0$.

We remark that if the predators consumed prey other than eggs, then a predation term would have to be included as a sink term on the right side of the dynamical equation (5.10).

The method of moments allows us to obtain a system of ordinary differential equations for the total prey and predator populations $N(t)$ and $P(t)$. In the analysis, we will also obtain equations for some additional auxiliary variables, but the overall end result is a system of ODEs, which is simpler than the mixed PDE-ODE system given by the model equations (5.10)–(5.13).

The idea is to multiply the PDE (5.10) by some "moment" function $g(a)$ and then integrate over $0 \le a \le \infty$. The only requirement is that $u(a, t)g(a) \to 0$ as $a \to \infty$. Upon taking g to be different functions, we can obtain equations that lead to the differential equations we seek. The reader will find it valuable to verify these calculations. Proceeding in a general way, we multiply the PDE by g and integrate to get

$$\frac{d}{dt}\int_0^\infty g(a)u(a, t)da = -\int_0^\infty g(a)u_a(a, t)da - m\int_0^\infty g(a)u(a, t)da.$$

The first integral on the right can be integrated by parts to get

$$\frac{d}{dt}\int_0^\infty g(a)u(a, t)da$$
$$= M(B, P)g(0) + \int_0^\infty g'(a)u(a, t)da - m\int_0^\infty g(a)u(a, t)da. \tag{5.14}$$

Now we make different choices for g. If $g(a) = 1$, then (5.14) becomes simply

$$\frac{dN}{dt} = M(B, P) - mN, \tag{5.15}$$

an ODE involving N, P, and B. If we take $g(a) = b(a)$, the maternity function, then (5.14) becomes

$$\frac{dB}{dt} = -\gamma B + b_0 H - mB, \tag{5.16}$$

where $H = H(t)$ is defined by

$$H(t) = \int_0^\infty e^{-\gamma a} u(a, t) da.$$

But now H is yet a new variable. To get another equation involving H we take $g(a) = e^{-\gamma a}$. Then (5.14) becomes

$$\frac{dH}{dt} = M(B, P) - (m + \gamma)H. \tag{5.17}$$

Therefore we have four ODEs [(5.13), (5.15), (5.16), (5.17)] for N, P, B, and H. Clearly the N equation decouples from the system and we can just consider the three ODEs

$$\frac{dP}{dt} = -\delta P + cBP, \quad \frac{dB}{dt} = -(m+\gamma)B + b_0 H, \quad \frac{dH}{dt} = M(B, P) - (m+\gamma)H.$$

The initial conditions are $P(0) = P_0$, $B(0) = \int_0^\infty b_0 a e^{-\gamma a} f(a) da$, $H(0) = \int_0^\infty e^{-\gamma a} f(a) da$.

We may now proceed with a numerical method (e.g., the Runge–Kutta method) to solve the system and determine the resulting dynamics. The reader is asked for such a calculation in Exercise 4.

Exercises

1. Consider a population of organisms whose *per capita* death rate is three percent per month and that the maternity function, in births per female per age in months, is given by $b(a) = 4$ for $3 \le a \le 8$, and $b(a) = 0$ otherwise. Use the Euler–Lotka equation to calculate the growth rate r.

2. Consider the age-structured model (5.1)–(5.3) in the case that the mortality rate is constant [$m(a) = \gamma$] and the maternity function is a constant [$b(a, t) = \beta$]. At time $t = 0$ assume the age distribution is $f(a) = u_0$ for $0 < a \le \delta$, and $f(a) = 0$ for $a > \delta$.

   (a) Show that the renewal equation (5.9) takes the form

   $$B(t) = \int_0^t \beta B(s) e^{-\gamma(t-s)} ds + \beta\delta u_0 e^{-\gamma t}.$$

   (b) Show that $B(t)$ satisfies the differential equation

   $$B' = (\beta - \gamma)B.$$

   (Hint: use Leibniz' rule to differentiate the integral.)

   (c) Determine $B(t)$ and the population density $u(a, t)$.

   (d) What is the total size $N(t)$ of the population at any time t?

3. Consider the structured model where the per capita mortality rate depends upon the total population $N = N(t)$ and the maternity function is $b(a) =$

$b_0 e^{-\gamma a}$:

$$u_t = -u_a - m(N)u, \quad a > 0, t > 0,$$

$$u(0,t) = \int_0^\infty b_0 e^{-\gamma a} u(a,t)da,$$

$$u(a,0) = f(a), \quad a > 0.$$

(a) Use the method of moments to obtain the system of ODEs

$$\frac{dN}{dt} = B - m(N)N, \quad \frac{dB}{dt} = (b_0 - \gamma - m(N))B,$$

for $N(t)$ and the offspring $B(t) = u(0,t)$. (Note that the maternity function in this model is unreasonable since it provides for newborns giving birth; but it may be a good approximation for the case when the population reproduces at a very young age.)

(b) Show that the relation $B = (b_0 - \gamma)N$ gives a solution to the ODEs in the NB-plane.

(c) Show that the solution to the system cannot oscillate and, in fact, approaches a steady state.

4. Numerically solve equations (5.13), (5.15), (5.16), (5.17) and plot the prey and predator populations $N(t)$ and $P(t)$ for $0 \le t \le 125$. Take $b_0 = 5$ and the remaining constants to be one. Based upon your calculation, is there a basis for controlling pests by introducing predators that selectively eat their eggs?

5. Consider an age-structured population where $u = u(a,t)$ is the age density of people at age a at time t. In a model where no person survives past age $a = d$, the PDE for u is given by

$$u_t + u_a = -\frac{c}{d-a}u, \quad 0 < a < d, \ t > 0.$$

Interpret this model and use `pdsolve` in Maple, or otherwise, to find the general solution. Then use analytical techniques to find the solution that satisfies the initial and boundary conditions

$$u(a,0) = f(a), \quad 0 \le a \le d; \quad u(0,t) = B(t), \quad t > 0,$$

where $f(a)$ is the initial age structure and $B(t)$ is a given birth schedule.

6. This exercise presents an alternate derivation of the McKendrick–von Foerster equation (5.1). If $u(a,t)$ is the density of females of age a at time t, then at a small time dt later, all females still alive will have aged an amount $da = dt$. Thus, the average number of females dying in this interval is

$$\frac{u(a+da, t+dt) - u(a,t)}{dt} = -m(a)u(a,t) + O(dt),$$

where $m(a)u(a,t)dt$ represents the probability of dying in the small time interval t to $t + dt$.

5.2 Traveling Waves Fronts

We observed in Chapter 1 that the advection equation $u_t = -cu_x$ and the wave equation $u_{tt} = -c^2u_{xx}$ both admit *traveling wave solutions* of the form $u(x, t) = U(z)$, where U is any function (the shape of the wave), and $z = x - ct$ is a moving coordinate that travels with speed c. As it turns out, many PDE models have these types of solutions, and we are interested in finding them. They model a signal, or disturbance, that moves with constant speed, and they can represent moving population fronts, infectious disease or epidemic waves, biological invasions, chemical reaction fronts, and many other interesting biological phenomena.

First consider the model diffusion-type equation having the general form

$$u_t = Du_{xx} + f(u, u_x), \tag{5.18}$$

where f is some fixed function, possibly nonlinear, depending the concentration u and its spatial gradient u_x. We are interested in finding solutions of the form $u(x, t) = U(z)$, where $z = x - ct$, where both the wave form U and the wave speed c are unknown. The spatial independent variable x is assumed to vary over all real numbers, that is, $-\infty < x < \infty$, and therefore $-\infty < z < \infty$. We can find an equation for the shape of the wave $U(z)$ by substituting this form into the PDE. First, we need to calculate the partial derivatives of u in order to make the substitution. Using the multi-variable chain rule, we get

$$\begin{aligned} u_t &= U'(z)z_t = -cU'(z), \\ u_x &= U'(z)z_x = U'(z), \\ u_{xx} &= U''(z)z_x = U''(z). \end{aligned}$$

Substituting into the PDE produces an ODE for the wave form $U(z)$:

$$-cU' = DU'' + f(U, U'), \quad -\infty < z < \infty. \tag{5.19}$$

Note that the ODE has only two variables, the independent "moving" coordinate z and the dependent wave form U. The variables t and x dropped out of the problem, which is required if the method is to succeed.

Ordinary differential equations are sometimes easier to deal with than PDEs. In the present case we have a second-order ODE where the independent variable is any real number. Solutions of such equations usually depend upon two arbitrary constants, and we need two auxiliary conditions to determine those constants. Therefore we impose conditions at $z = \pm\infty$ on the wave form, namely that $U(z)$ approach constant values, or densities, at both extremes:

$$U(-\infty) = u_l, \quad U(+\infty) = u_r, \tag{5.20}$$

where u_l and u_r are two fixed constants. We assume that the these two states are equilibrium states for the equation, i.e., $f(u_l, 0) = f(u_r, 0) = 0$.

Thus we are not interested in any traveling wave, but only those that are bounded and approach constant, equilibrium states at $z = \pm\infty$. These special types of traveling wave solutions are called *wave front solutions*. See figure 5.3.

To complicate matters, the wave speed c is also unknown and often acts as an "eigenvalue"; that is, wave front solutions will exist only for certain values of c. Hence, we often refer to (5.19)–(5.20) as a nonlinear eigenvalue problem.

Wave fronts can arise from natural settings as limiting, or long time, solutions to boundary value problems. As an illustration, consider the initial BVP for a toxic chemical of concentration $u = u(x, t)$ in a semi-infinite domain, e.g., a long canal:

$$u_t = Du_{xx} + f(u, u_x), \quad x > 0, \quad t > 0,$$

$$u(x, 0) = 0, \quad x > 0,$$

$$u(0, t) = 1, \quad t > 0,$$

$$u(\infty, t) = 0.$$

Here f is a term that contains advection and reaction processes. Intuitively, we reason as follows. For early times a concentration wave with unit magnitude at the left boundary ($x = 0$) begins to move into the medium; at these times the shape of the wave changes as it evolves. However, over a longer time the concentration wave can begin to approach a wave front with the same shape wave form. That is, the solution to the initial BVP may approach a wave front having the form $u = F(x - ct)$ for some wave speed c. Often we just look for the wave front. The issue of whether solutions to initial BVPs evolve into wave fronts is a question of whether the wave front attracts other solutions, i.e., it is one of stability of the front.

Generally we are not able to find an explicit solution to (5.19) and we must be content to find a numerical solution. Oftentimes we can prove there must be a unique solution without solving the equation in any manner. A standard technique is to transform (5.19) to a system of first-order ODEs for the variables U and $V = U'$ in the UV phase plane. We

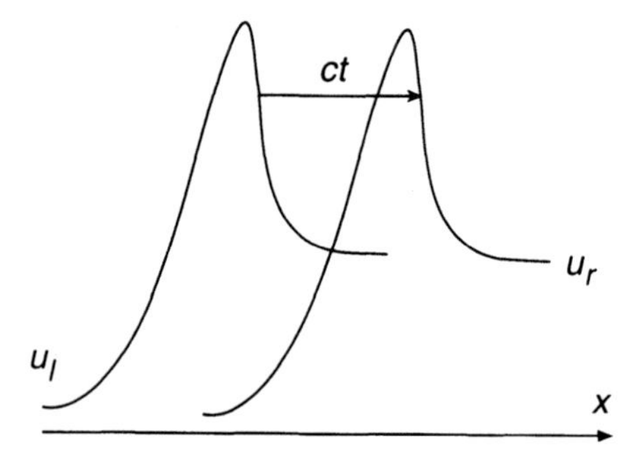

Figure 5.3. Plot of two successive wave profiles of a wave front solution.

obtain

$$U' = V,$$

$$V' = -\frac{c}{D}U' - \frac{1}{D}f(U, U').$$

Assuming that V, or equivalently U', goes to zero at $z = \pm\infty$, the problem of finding a wave front solution to (5.19) now consists of finding a solution (trajectory) in phase space that connects the two equilibrium points $(u_l, 0)$ and $(u_r, 0)$. A thorough discussion of the phase space method is beyond our scope and we refer the reader to other sources [e.g., see Logan (1994, 2001), or the references given at the beginning of this chapter.]

EXAMPLE
Consider the diffusion equation

$$u_t = Du_{xx}.$$

Solutions of the form $u(x, t) = U(z)$, where $z = x - ct$, must satisfy the ODE

$$-cU' = DU''.$$

This is a linear second-order equation with characteristic equation (see Appendix A) $-cm = Dm^2$ having roots $m = 0$, $m = -c/D$. Therefore, the independent solutions are 1 and $e^{-cz/D}$, and the general solution is therefore

$$U(z) = c_1 + c_2 e^{-cz/D},$$

or

$$u(x, t) = c_1 + c_2 e^{-c(x-ct)/D}.$$

Therefore, the diffusion equation admits traveling wave solutions, but not (nonconstant) wave front solutions because $e^{-cz/D}$ does not remain bounded at $z = +\infty$. □

EXAMPLE
(Disease model) We consider a model for the spread of a rabies epidemic throughout a healthy population of foxes. We set up the problem in one-dimensional linear geometry, and we ignore natural mortality and births. Let $S = S(x, t)$ denote the density of the susceptible fox population and $I = I(x, t)$ the density of the infected foxes. We assume the rate that healthy foxes become infected is proportional to SI, the argument being that the number of contacts between members of the two groups is equal to their product; thus, the rate of conversion of susceptibles to infecteds is bSR, where b is the constant infection rate, or the fraction of the contacts that lead to infection. Because rabies is fatal, we assume that infected foxes die at *per capita* rate r and they are removed from the susceptible and infected populations. Finally, we assume that healthy foxes are territorial

and do not diffuse, while rabid foxes are disoriented and diffuse randomly with diffusion constant D. Thus we have the model equations

$$S_t = -bSI,$$
$$I_t = bSI - rI + DI_{xx}.$$

By an epidemic wave we mean a wave of infecteds moving into a healthy population, and we model it by a wave front of unknown wave speed c. Letting $S = S(z)$, $I = I(z)$, $z = x - ct$, we get wave front equations

$$-cS' = -bSI,$$
$$-cI' = bSI - rI + DI''.$$

At $z = +\infty$, ahead of the infection or epidemic wave, we assume there are only susceptibles, so $S(+\infty) = N$, where N is the total population of foxes, and $I(+\infty) = 0$. After the wave has passed, at $z = -\infty$, we assume no infecteds remain, or $I(-\infty) = 0$ (the epidemic has died out), and the number of susceptibles is $S(-\infty) = S_l$, which is not known *a priori*. We want to determine the speed c of the epidemic wave and S_l, the number that will not get the disease. Figure 5.4 shows the shapes of the waves that we might expect.

To make progress, note that $S'/S = (\ln S)'$, so the first equation becomes $I = (c/b)(\ln S)'$. We can substitute this into the second equation to obtain

$$-cI' = cS' - \frac{rc}{b}(\ln S)' + DI'',$$

which can be integrated with respect to z since every term is a derivative. Then

$$-cI = cS - \frac{rc}{b}\ln S + DI' + a,$$

where a is a constant of integration. Taking the limit as $z \to +\infty$ gives $a = c(\frac{r}{b}\ln N - N)$. Taking the limit as $z \to -\infty$ gives another equation relating a and c, namely $a = c(\frac{r}{b}\ln S_l - S_l)$. Therefore,

$$\frac{r}{b}\ln N - N = \frac{r}{b}\ln S_l - S_l, \tag{5.21}$$

which is a relation between the number S_l that do not get the disease, the total initial population N, the infection rate b, and the death rate r. The relation is independent of the epidemic wave speed c. To simplify,

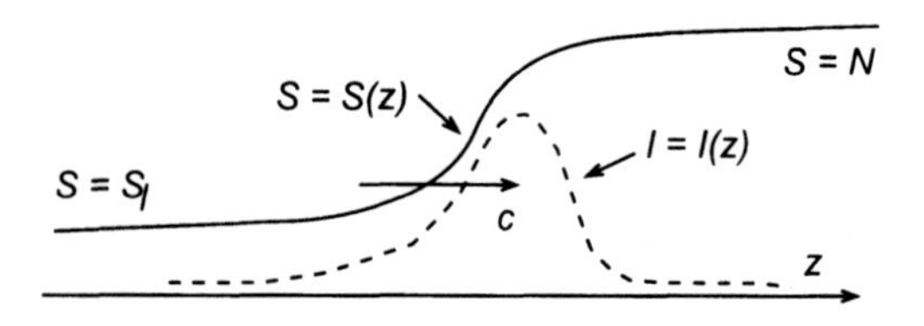

Figure 5.4. *SI* epidemic wave.

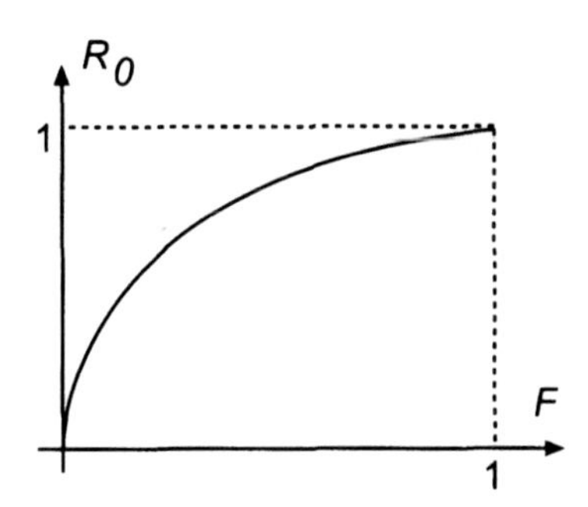

Figure 5.5. Plot of F vs R_0.

we introduce parameters

$$R_0 = \frac{r}{Nb}, \quad F = \frac{S_l}{N},$$

where F is the fraction that do not contract rabies and R_0 is the reproductive ratio of the disease, representing the death rate divided by the infection rate. Then (5.21) can be written in the form

$$R_0 = \frac{F - 1}{\ln F}, \quad 0 < F < 1.$$

A plot of this function is shown in figure 5.5, from which we infer that R_0 cannot exceed 1. Thus, no epidemic wave front can occur when $R_0 > 1$. This is reasonable since the death rate would be high compared to the rate of getting the disease; the epidemic dies out because the infecteds die out faster than they are replaced. This case can occur if b is small or can be forced small, say, by inoculation. On the other hand, if $R_0 < 1$, then a epidemic wave can propagate; the smaller R_0, the greater the number that get rabies. □

We have not solved the wave front equations to find the wave forms or even shown that such wave forms exists; nor have we determined the speed of the epidemic. The goal has been to derive certain properties of the epidemic wave under the assumption that it exists. More thorough analyses, given in the references, show that such solutions do exist.

Exercises

1. Show that the nonlinear advection–diffusion equation

$$u_t = Du_{xx} - uu_x$$

admits a wave front solution $U = U(z)$, satisfying the boundary conditions $U(-\infty) = 1$, $U(+\infty) = 0$. Find the wave speed and sketch the the solution on a zu-coordinate system if $U(0) = \frac{1}{2}$.

2. Consider a reaction–diffusion equation with a nonlinear advection term given by

$$u_t = u_{xx} - u^2 u_x.$$

For which wave speeds c does a positive wave front solution exist with $u \to 0$ as $x \to \infty$? What are the possible states at $x = -\infty$?

3. Consider the system of reaction–advection–diffusion equations

$$u_t = Du_{xx} - \gamma u_x - aF(u, v),$$
$$v_t = \quad - bF(u, v),$$

where a, b, D, and γ are positive constants. Find a system of ODEs for traveling wave forms $u = U(z)$, $v = V(z)$, $z = x - ct$. If boundary conditions are given by

$$U(-\infty) = u_l, \quad U(+\infty) = 0, \quad V(-\infty) = 0, \quad V(+\infty) = v_r,$$

what conditions must the reaction term F satisfy for wave front solutions to exist? Sketch possible wave front profiles. Assuming wave fronts exist, show that the speed c of the wave is less than the advection speed γ.

4. The following system of PDEs arises in the study of bioremediation of aquifer systems where immobile, indigenous microbes attached to the soil are stimulated to consume a contaminant and produce nontoxic products (see Logan, 2001, p107):

$$RS_t = -vS_x - F, \quad A_t = -vA_x - rF, \quad M_t = yF - b(M - M_0),$$

where $M(x, t)$ is the density of the microbes, $S(x, t)$ is the density of the contaminant (e.g., a hydrocarbon), $A(x, t)$ is the density of a nutrient stimulant (e.g., oxygen), and $F = qSAM/[(K_s + S)(K_a + A)]$ is the biodegradation rate. $R > 1$ is the retardation constant, v is the average velocity of the subsurface flow, b is the decay rate of the bacteria, y is the yield, M_0 is a reference microbe density, and r is the mass of the nutrient used per mass contaminant degraded; q, K_s, and K_a are rate constants. Find the speed c of an assumed wave front that satisfies the boundary conditions $S = 0$, $A = A_0$ at $-\infty$, $S = S_r$, $A = 0$ at $+\infty$, and $M = M_0$ at $\pm\infty$. Without solving the wave front differential equations, sketch anticipated profiles of S, A, and M as a function of the variable $z = x - ct$.

5. The nonlinear PDE model

$$((1 + b)u - mu^2)_t = u_{xx} - u_x$$

arises in the subsurface transport of a contaminant that is adsorbed to the soil. Assume $b > m$. Find a wave front solution $u = U(x - ct)$ satisfying the conditions $U(-\infty) = 1$, $U(+\infty) = 0$, $U(0) = 0.5$. The solution is

$$u(x, t) = \frac{1}{1 + e^{cm(x-ct)}}.$$

5.3 Equilibria and Stability

We already noted the broad occurrence of diffusion problems in biological systems. Now we investigate another aspect of such problems, namely the persistence of equilibrium states in systems governed by reaction–diffusion systems. At issue is the stability of those states: if a system is in an equilibrium state and it is slightly perturbed, or displaced, does the system return to that state, or does it evolve to a completely different state?

Underpinned by the seminal work of Alan Turing in 1952 on the chemical basis of morphogenesis, it has been shown in extensive research that diffusion-induced instabilities can give rise to spatial patterns in all sorts of biological systems. Reaction–diffusion models have been used to explain the evolution of form and structure in developmental biology (morphogenesis), tumor growth, ecological spatial patterns, aggregation, patterns on animal coats, and many other processes in molecular and cellular biology.

In this section we introduce the basic idea of stability in reaction–diffusion models and we observe that such systems can have instabilities that lead to density variations and patterns. We refer the reader to Murray (2003) for an in-depth treatment of these models in the life sciences.

Stability for ODEs

To illustrate the notion of stability of an equilibrium, or steady-state, to a PDE, we begin with a review of the stability issue for ODEs. For example, consider the logistics population model

$$\frac{du(t)}{dt} = ru(t)\left(1 - \frac{u(t)}{K}\right), \tag{5.22}$$

where r is the intrinsic growth rate and K is the carrying capacity, both positive. A steady-state, or *equilibrium solution*, is a constant solution $u = u_e$. For such a solution the left side of (5.22) must be zero, and so u_e satisfies the algebraic equation

$$ru_e\left(1 - \frac{u_e}{K}\right) = 0.$$

Therefore there are two equilibria, $u_e = 0$ and $u_e = K$. If at some time t the system is in an equilibrium state, it remains in that state for all time. That, if $u(t_0) = u_e$ for some t_0, then $u(t) = u_e$ for all $t > t_0$ because $u(t) = u_e$ satisfies both the initial condition and the differential equation; uniqueness implies this. Natural "perturbations," however, often disturb equilibrium and move the system a small amount away from its steady state. Does the system return to that state, or does it deviate significantly from that state, possible going to another equilibrium state? This is the question of stability. In the present example, if the system is at carrying capacity $u = K$ and a small number of organisms are removed, does the system grow back to that original state or does it do something else? To answer this question we try to determine how a small deviation, or *perturbation*, will evolve by finding an equation for the deviation and solving it. Therefore, let $U(t)$ be a small deviation from the steady state $u = K$. That is, assume $u(t) = K + U(t)$. Then $u(t)$ must still solve the model equation. Substituting into (5.22) gives

$$\frac{d(K + U(t))}{dt} = r(K + U(t))\left(1 - \frac{K + U(t)}{K}\right).$$

Simplifying the right side and noting that $dK/dt = 0$ gives

$$\frac{dU(t)}{dt} = -\frac{r}{K}(KU(t) + U(t)^2), \tag{5.23}$$

which is called the *perturbation equation*. Immediately we observe that the right side is negative, which means that $dU/dt < 0$. Consequently, the perturbation must decay away regardless of its initial value. This means that the equilibrium state $u = K$ is stable to small perturbations; the system returns to the equilibrium state.

Note that we did not solve the perturbation equation (5.23) to make this stability assertion. We could have solved it as a Bernoulli equation (Appendix A), but it was unnecessary. In general we cannot solve the perturbation equation because it is usually nonlinear. But there is another argument we can use, called *linearization*, to make conclusions about stability. In (5.23) the perturbations are small so that the $U(t)^2$ term is very small compared to the $U(t)$ term. If we discard the nonlinear $U(t)^2$ term, we obtain a *linearized perturbation equation*

$$\frac{dU(t)}{dt} = -rU(t).$$

The equation can be solved instantly to obtain $U(t) = Ce^{-rt}$, which shows that the perturbations decay.

This linearization procedure can be applied to any nonlinear autonomous equation

$$\frac{du}{dt} = f(u), \quad u = u(t).$$

Equilibrium solutions $u = u_e$ are solutions of the algebraic equation

$$f(u_e) = 0. \tag{5.24}$$

If $u(t) = u_e + U(t)$, where $U = U(t)$ is a small deviation from equilibrium, then it satisfies the perturbation equation

$$\frac{dU}{dt} = f(u_e + U).$$

Since U is assumed small, we may expand the right side in its Taylor series about u_e. Thus,

$$f(u_e + U) = f(u_e) + f'(u_e)U + \frac{1}{2}f''(u_e)U^2 + \cdots.$$

Discarding the higher-order terms, while using (5.29), we obtain a linearized perturbation equation

$$\frac{dU}{dt} = f'(u_e)U.$$

This equation has general solution $U(t) = Ce^{\lambda t}$, where $\lambda = f'(u_e)$ is termed the *eigenvalue*; if $\lambda < 0$, the perturbation decays and the equilibrium u_e is stable, and if $\lambda > 0$, the perturbation grows and the equilibrium is unstable. If $\lambda = 0$, there is no information and we have to take into account additional terms in the Taylor series. In summary, the stability of a equilibrium state can be determined by examining the sign of the eigenvalue, i.e., the sign of the derivative of the right side of the differential equation, evaluated at that equilibrium state.

The linearization procedure is based upon the assumption that the perturbations are sufficiently small. The type of stability that results is called *local stability* since the deviations are not far from the equilibrium state. The procedure gives no results about *global stability*, i.e., how the system would evolve if any perturbation, regardless of its magnitude, were permitted.

EXAMPLE

In the logistics equation (5.22) the right side is

$$f(u) = ru\left(1 - \frac{u}{K}\right).$$

Therefore,

$$f'(u) = r - \frac{2r}{K}u.$$

To determine local stability of the equilibrium state $u_e = K$ we check the eigenvalue

$$\lambda = f'(K) = r - \frac{2r}{K}K = -r.$$

Because $\lambda < 0$ the equilibrium $u_e = K$ is stable; small perturbations decay away. The other equilibrium solution is $u_e = 0$. In this case $\lambda = f'(0) = r > 0$, and therefore the zero state is unstable; if an initial small deviation were imposed, it would grow and the system would depart from equilibrium. □

Stability for PDEs

A similar theory of local stability of equilibrium solutions can be developed for PDEs. In the interval $0 < x < L$ consider the reaction–diffusion equation

$$u_t = Du_{xx} + f(u),$$

with no-flux boundary conditions $u_x(0, t) = u_x(\pi, t) = 0$. Let $u(x, t) = u_e$ be a constant equilibrium solution [so that $f(u_e) = 0$]. To fix the idea let $a = f'(u_e) > 0$. Note that the equilibrium solution satisfies the PDE and the boundary conditions. Next, let $U(x, t)$ be a small perturbation from the equilibrium solution, or $u(x, t) = u_e + U(x, t)$. The *perturbation equation* is determined by substituting this expression into the PDE and the boundary conditions:

$$U_t = DU_{xx} + f(u_e + U), \quad U_x(0, t) = U_x(L, t) = 0.$$

To linearize this equation we expand, as above, the right side in a Taylor series and discard the nonlinear terms:

$$f(u_e + U) = f(u_e) + f'(u_e)U + \cdots = f'(u_e)U + \cdots .$$

Then we obtain the *linearized perturbation equation*

$$U_t = DU_{xx} + aU, \qquad a = f'(u_e) > 0 \tag{5.25}$$

subject to boundary conditions

$$U_x(0, t) = U_x(L, t) = 0. \tag{5.26}$$

This problem, which is on a bounded interval, can be solved by the separation of variables method (Chapter 4). We assume $U = g(x)h(t)$ and then substitute into the PDE (5.25) and boundary conditions (5.26) to obtain

$$gh' = Dg''h + agh,$$

or

$$\frac{h'}{h} = \frac{Dg''}{g} + a = \lambda.$$

Then $h = Ce^{\lambda t}$ and

$$g'' + \left(\frac{a - \lambda}{D}\right) g = 0, \quad g'(0) = g'(L) = 0.$$

This equation cannot have nontrivial exponential solutions that satisfy the boundary conditions; therefore, $a - \lambda \geq 0$. In the case $\lambda = a$ we obtain a constant solution, and in the case $a - \lambda > 0$ we obtain

$$g(x) = A \sin \sqrt{\frac{a - \lambda}{D}}\, x + B \cos \sqrt{\frac{a - \lambda}{D}}\, x.$$

Applying $g'(0) = 0$ forces $A = 0$; then $g'(L) = 0$ implies

$$\sin \sqrt{\frac{a - \lambda}{D}}\, L = 0,$$

or

$$\sqrt{\frac{a-\lambda}{D}}L = n\pi, \quad n = 1, 2, \ldots .$$

Therefore, incorporating the case $\lambda = a$, we obtain eigenvalues

$$\lambda = \lambda_n = a - \frac{Dn^2\pi^2}{L^2}, \quad n = 0, 1, 2, \ldots . \tag{5.27}$$

The modal solutions are therefore

$$U_n(x, t) = e^{\lambda_n t} \cos \frac{n\pi}{L} x, \quad n = 0, 1, 2, \ldots . \tag{5.28}$$

What conclusions can we draw from this calculation? Because the general solution $U(x, t)$ to the boundary value problem (5.25)–(5.26) is a linear combination of the modal solutions (5.28),

$$U(x, t) = c_0 e^{at} + \sum_{n=1}^{\infty} c_n e^{\lambda_n t} \cos \frac{n\pi}{L} x,$$

it will decay if all of the modes decay, and it will grow if one of the modes grows. The constants c_n are determined by the initial perturbation. The spatial part of the modal solutions, i.e., the cosine, remains bounded. The amplitude factor $e^{\lambda_n t}$, and therefore the eigenvalues λ_n, determine growth or decay. Let us examine the modes. In the case $n = 0$ the eigenvalue is $\lambda_n = a$ and the modal solution is $c_0 e^{at}$, which grows exponentially. In fact, any mode satisfying $a \geq \frac{Dn^2\pi^2}{L^2}$ is unstable. Therefore, instabilities are likely to occur for modes of low frequency (n small), or in systems of large size L or low diffusion properties D. Oppositely, small systems with large diffusion constants are stabilizing, as are the high-frequency modes. In the most general case the initial perturbation will contain all modes and the steady state will be locally unstable.

Cell aggregation

We apply these ideas to a problem in cell aggregation. A slime mold population is a collection of unicellular ameboid cells that feed on bacteria in the soil. When the food supply is plentiful, the bacteria are generally uniformly spatially distributed throughout the soil; but as the food supply becomes depleted and starvation begins, the ameba start to secrete a chemical (cyclic AMP) that acts as an attractant to the other ameba and aggregation sites form. The rest of the story is even more interesting as the aggregation sites evolve into slugs that ultimately develop into a sporangiophores consisting of a stalk and head containing new spores. The spores are released and the process begins anew. We are interested here in only the first part of this complicated problem, the onset of aggregation. We work in one spatial dimension.

Let $a = a(x, t)$ and $c = c(x, t)$ denote the density and concentration of the cellular ameba and cyclic AMP, respectively. The fundamental conservation laws are

$$a_t = -\phi_x^{(a)}, \quad c_t = -\phi_x^{(c)} + F,$$

where $\phi^{(a)}$ and $\phi^{(c)}$ are the fluxes of the ameba and the chemical, respectively. There are no source terms in the ameba equation because we will not consider birth and death processes on the time scale of the analysis. The source term in the chemical equation consists of two parts: production by the ameba and degradation in the soil. We assume the chemical is produced by the ameba at a rate proportional to the density of the ameba, and the chemical degrades at a rate proportional to its concentration; i.e.,

$$F = fa - kc.$$

The chemical moves by diffusion only, and we assume Fick's law:

$$\phi^{(c)} = -\delta c_x,$$

where δ is the diffusion constant. The ameba are also assumed to randomly diffuse, but there is another flux source for the ameba, namely attraction to the chemical. We assume this attraction is "up the chemical gradient," toward high concentrations of c. Additionally, it should depend on the ameba population since that will increase the magnitude of the chemical concentration released. This type of flow, induced by chemical gradients, is called *chemotaxis*. Therefore we assume

$$\begin{aligned} \phi^{(a)} &= \text{random flux} + \text{chemotatic flux} \\ &= -\mu a_x + \nu a c_x, \end{aligned}$$

where μ is the ameba motility and ν is the strength of the chemotaxis, both assumed to be positive constants. Note that the random flux, having the form of Fick's law, has a negative sign since flow is "down the gradient" (from high to low densities), and the chemotatic flux has a positive sign since that term induces flow "up the chemical gradient" (from low to high concentrations). Putting all these equations all together gives

$$a_t = \mu a_{xx} - \nu (a c_x)_x, \quad c_t = \delta c_{xx} + fa - kc. \tag{5.29}$$

Both a and c satisfy no-flux boundary conditions, i.e., $a_x = c_x = 0$ at $x = 0, L$, which means there is no escape from the medium.

Note that there will be an equilibrium solution $a = \bar{a}$, $c = \bar{c}$ to (5.29) provided

$$f\bar{a} = k\bar{c}.$$

That is, the production of the chemical equals its degradation. This equilibrium state represents the spatially uniform state in the soil before aggregation begins.

To determine the local stability of this state we let

$$a = \bar{a} + A(x,t), \quad c = \bar{c} + C(x,t),$$

where A and C are small perturbations. Substituting these quantities into (5.29) gives, after simplification, the perturbation equations

$$A_t = \mu A_{xx} - \nu((\overline{a} + A)C_x)_x, \quad C_t = \delta C_{xx} + fA - kC.$$

These equations are nonlinear because of the AC_x term in the ameba equation. If we discard the nonlinear term on the assumption that the product of small terms is even smaller, then we obtain the linearized perturbation equations

$$A_t = \mu A_{xx} - \nu\overline{a}C_{xx}, \quad C_t = \delta C_{xx} + fA - kC. \tag{5.30}$$

Easily one can see that the perturbations satisfy the no-flux boundary conditions.

Motivated by our knowledge of the form of solutions to linear equations, we assume there are modal solutions of the form

$$A(x,t) = c_1 e^{\sigma t} \cos rx, \quad C(x,t) = c_2 e^{\sigma t} \cos rx, \tag{5.31}$$

where r and σ are to be determined, and c_1 and c_2 are some constants. Note the form of these assumed solutions. The spatial part is bounded and periodic with frequency r and period $2\pi/r$, and the temporal part is exponential with growth factor σ, which may be a real or complex number. If σ is negative or has negative real part, then the perturbation will decay and the equilibrium state will return (stable); if σ is positive or has positive real part, then the perturbations will grow and the equilibrium will be unstable. Let us ask why solutions of (5.30) should be of this form (5.31) without going through the entire separation of variables argument. Equations (5.30) are linear with constant coefficients, and both equations contain both unknowns A and C and their derivatives. If we are to have a solution, then all the terms must match up in one way or another in order to cancel. So both A and C must essentially have the same form. Because there is a first derivative in t, the time factor must be exponential so it will cancel, and the spatial part must be a sine or a cosine because of the appearance of a second spatial derivative. We anticipate the cosine function because of the no-flux boundary conditions, as in Chapter 4. If we substitute (5.31) into (5.30) we obtain the two equations

$$(\sigma + \mu r^2)c_1 - \nu\overline{a}r^2 c_2 = 0, \quad -fc_1 + (\sigma + k + \delta r^2)c_2 = 0$$

which relate all the parameters. We may regard these two equations as two linear, homogeneous for the constants c_1 and c_2. If we want a nontrivial solution for c_1 and c_2, then matrix theory dictates that the determinant of the coefficient matrix must be zero. That is,

$$(\sigma + \mu r^2)(\sigma + k + \delta r^2) - f\nu\overline{a}r^2 = 0,$$

which is an equation relating the temporal growth factor σ, the spatial frequency r, and the other constants in the problem. Expanded out, this equation is a quadratic in σ,

$$\sigma^2 + \gamma_1\sigma + \gamma_2 = 0,$$

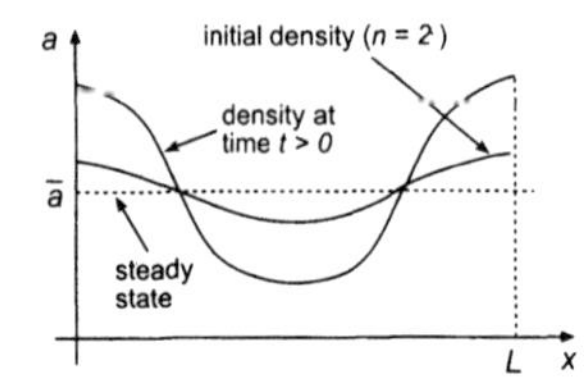

Figure 5.6. Plot showing the growing ameba density at time t when the uniform state is unstable to local perturbations in the mode $n = 2$.

where

$$\gamma_1 = r^2(\mu + \gamma) + k > 0, \quad \gamma_2 = r^2[\mu(\delta r^2 + k) - f\nu\bar{a}].$$

The roots of the quadratic are

$$\sigma = \frac{1}{2}\left(-\gamma_1 \pm \sqrt{\gamma_1^2 - 4\gamma_2}\right).$$

Clearly one of the roots is always negative or has negative real part. The other root can have positive or negative real part, depending upon the value of the discriminant $\gamma_1^2 - 4\gamma_2$. We are interested in determining if there are parameter choices that lead to an instability; so we want σ positive. Hence, γ_2 must be negative, or

$$\mu(\delta r^2 + k) < f\nu\bar{a}.$$

If this inequality holds, there is an unstable mode and perturbations will grow. We analyze this further.

The number r is the frequency of the perturbations. Applying the no-flux boundary conditions forces

$$r = \frac{n\pi}{L}, \quad n = 0, 1, 2, 3, \ldots.$$

For each value of n we obtain a frequency $r = r_n$ and a corresponding growth factor σ_n. The nth mode will therefore grow and lead to local instability when

$$\mu\left(\delta\frac{n^2\pi^2}{L^2} + k\right) < f\nu\bar{a}. \tag{5.32}$$

We can now ask what factors destabilize the uniform, equilibrium state in the ameba–cAMP system and therefore promote aggregation. That is, when is (5.32) likely to hold? We can list the factors that may make the left side of the inequality (5.32) smaller than the right side: low motility μ of the bacteria; low degradation rate k or large production rate f of cAMP; large chemotactic strength ν; large dimensions L of the medium; a small value of n (thus, low-frequency, or long-wavelength, perturbations are less stabilizing that short-wavelength, or high-frequency, perturbations); decreasing the diffusion constant of the cAMP. Figure 5.6 shows time snapshots of the ameba density for the mode $n = 2$ when it is unstable. The regions where the amplitude is high corresponds to higher concentrations of ameba, i.e., regions of aggregation.

Exercises

1. Consider a fish population that grows logistically, and at the same time is harvested at a rate proportional to the population. The model is

$$\frac{du}{dt} = ru\left(1 - \frac{u}{K}\right) - hu$$

where r, K, and h are the growth rate, carrying capacity, and harvesting rate, respectively. Find all the equilibria and analyze their stability. What are the possible long-term fish populations?

2. (Turing system) Consider the system of reaction diffusion equations on the spatial domain $0 < x < L$ given by

$$u_t = \alpha u_{xx} + f(u, v), \quad v_t = \beta v_{xx} + g(u, v)$$

with no-flux boundary conditions $u_x = v_x = 0$ at $x = 0, L$. Let $u = \overline{u}, v = \overline{v}$ be an equilibrium solution and define small perturbations U and V from equilibrium given by

$$u = \overline{u} + U(x, t), v = \overline{v} + V(x, t).$$

(a) Show that U and V satisfy no-flux boundary conditions and the linearized perturbation equations

$$U_t = \alpha U_{xx} + f_u(\overline{u}, \overline{v})U + f_v(\overline{u}, \overline{v})V, \tag{5.33}$$

$$V_t = \beta V_{xx} + g_u(\overline{u}, \overline{v})U + g_v(\overline{u}, \overline{v})V. \tag{5.34}$$

(b) Introduce matrix notation

$$\vec{W} = \begin{pmatrix} U \\ V \end{pmatrix}, \quad D = \begin{pmatrix} \alpha & 0 \\ 0 & \beta \end{pmatrix}, \quad J = \begin{pmatrix} f_u(\overline{u}, \overline{v}) & f_v(\overline{u}, \overline{v}) \\ g_u(\overline{u}, \overline{v}) & g_v(\overline{u}, \overline{v}) \end{pmatrix}$$

and show that (5.33)–(5.34) can be written as

$$\vec{W}_t = D\vec{W}_{xx} + J\vec{W}. \tag{5.35}$$

(c) Assume modal solutions to (5.35) of the form

$$\vec{W} = \vec{C} e^{\sigma_n t} \cos\frac{n\pi x}{L}, \qquad \vec{C} = \begin{pmatrix} c_{1n} \\ c_{2n} \end{pmatrix}, \quad n = 0, 1, 2, \ldots,$$

and show that for a nontrivial solution we must have

$$\det\left(\sigma_n I + \frac{n^2\pi^2}{L^2} D - J\right) = 0. \tag{5.36}$$

(When expanded, this equation is a quadratic equation for the growth factor σ_n of the nth mode. The roots σ_n depend upon the diffusion constants α, β, the equilibrium solution $\overline{u}, \overline{v}$, the size of the medium L, and the wavelength $2L/n$ of the perturbation. If one can find values of the parameters that make one of the roots positive or have positive real part, then there is an unstable mode.)

3. Apply the method of Exercise 2 to examine the stability of the steady state of the Turing system

$$u_t = Du_{xx} + 1 - u + u^2 v, \quad v_t = v_{xx} + 2 - u^2 v, \quad 0 < x < \pi,$$

under no-flux boundary conditions. Specifically, write the condition (5.36) and determine values of D for which various modes (n) are unstable.

4. Consider Fisher's equation with Dirichlet boundary conditions:

$$u_t = u_{xx} + u(1-u), \quad -\frac{\pi}{2} < x < \frac{\pi}{2},$$

$$u = 3 \quad \text{at} \quad x = \pm\frac{\pi}{2}.$$

(a) Show that $u_e(x) = \frac{3}{1+\cos x}$ is a *nonconstant* steady-state solution.

(b) Define perturbations $U(x,t)$ by the equation $u = u_e(x) + U(x,t)$ and find the linearized perturbation equation and boundary conditions for $U(x,t)$.

(c) Assume a solution to the linearized equation of the form $U = e^{\sigma t} g(x)$ and show that g must satisfy

$$g'' + \frac{\cos x - 5}{1 + \cos x} g = \sigma g, \quad g = 0 \;\text{ at }\; x = \pm\frac{\pi}{2}. \tag{5.37}$$

(d) Show that if (5.37) has a nontrivial solution, then $\sigma < 0$, thereby showing local stability of the steady solution. Hint. Consider two cases, when g is positive and when g is negative on the interval and then examine at the signs of g'' and the other terms in (5.37) at a maximum or minimum point.

APPENDIX A Ordinary Differential Equations

PDE models are frequently solved by reducing them to one or more ODEs. This appendix contains a brief review of how to solve some of the basic ODEs that are encountered in this book. At the end of the appendix are several exercises that should be solved *by hand*; the reader might want to check the solutions using a computer algebra package.

For notation, we let $y = y(x)$ be the unknown function. Derivatives will be denoted by primes, i.e., $y' = y'(x)$, $y'' = y''(x)$. Sometimes we use the differential notation $y' = \frac{dy}{dx}$. If f is a function, an *antiderivative* is defined as a function F whose derivative is f, i.e., $F'(x) = f(x)$. Antiderivatives are unique only up to an additive constant, and they are often denoted by the usual indefinite integral sign:

$$F(x) = \int f(x)dx + C.$$

An arbitrary constant of integration C is added to the right side. However, in this last expression, it is sometimes impossible to evaluate the antiderivative F at a particular value of x. For example, if $f(x) = \sin x/x$, then there is no simple formula for the antiderivative; that is,

$$F(x) = \int \frac{\sin x}{x} dx$$

cannot be expressed in closed form in terms of elementary functions, and thus we could not find, for example, $F(2)$. Therefore, it is better to denote the antiderivative by an integral with a variable upper limit,

$$F(x) = \int_a^x f(s)ds + C,$$

where a is any constant (observe that a and C are not independent, since changing one changes the other). By the fundamental theorem of calculus, $F'(x) = f(x)$. Now, for example, the antiderivative of $\sin x/x$ can be written (taking $C = 0$)

$$F(x) = \int_0^x \frac{\sin s}{s}\, ds,$$

and easily we find that

$$F(2) = \int_0^2 \frac{\sin s}{s}\, ds \approx 1.605.$$

First-Order Equations

An ODE of the first order is an equation of the form

$$G(x, y, y') = 0.$$

There are three types of these equations that occur regularly in PDEs: separable, linear, and Bernoulli. The general solution involves an arbitrary constant C that can be determined by an initial condition of the form $y(x_0) = y_0$.

Separable Equations

A first-order equation is separable if it can be written in the form

$$\frac{dy}{dx} = f(x)g(y).$$

In this case we separate variables to write

$$\frac{dy}{g(y)} = f(x)dx.$$

Then we can integrate both sides to get

$$\int \frac{dy}{g(y)} = \int f(x)dx + C,$$

which defines the solution implicitly. As noted above, sometimes the antiderivatives should be written as definite integrals with a variable upper limit of integration.

The simplest separable equation is the growth–decay equation

$$y' = \lambda y,$$

which has general solution

$$y = ce^{\lambda x}.$$

The solution models exponential growth if $\lambda > 0$ and exponential decay if $\lambda < 0$.

Linear Equations

A first-order linear equation is one of the form

$$y' + p(x)y = q(x).$$

This can be solved by multiplying through by an integrating factor of the form

$$e^{\int_a^x p(s)ds}.$$

This turns the left side of the equation into a total derivative, and it becomes

$$\frac{d}{dx}\left(y\exp(\int_a^x p(s)ds)\right) = q(x)\exp(\int_a^x p(s)ds)$$

Now, both sides can be integrated from a to x to find y. We illustrate this procedure with an example.

EXAMPLE

Find an expression for the solution to the initial value problem

$$y' + 2xy = \sqrt{x}, \quad y(0) = 3.$$

The integrating factor is $\exp(\int_0^x 2sds) = \exp(x^2)$. Multiplying both sides of the equation by the integrating factor gives

$$(ye^{x^2})' = \sqrt{x}e^{x^2}.$$

Now, integrating from 0 to x (while changing the dummy variable of integration to s) gives

$$y(x)e^{x^2} - y(0) = \int_0^x \sqrt{s}e^{s^2}ds.$$

Solving for y gives

$$y(x) = e^{-x^2}(3 + \int_0^x \sqrt{s}e^{s^2}ds) = 3e^{-x^2} + \int_0^x \sqrt{s}e^{s^2-x^2}ds.$$

As is frequently the case, the integrals in this example cannot be computed easily, if at all, and we must write the solution in terms of integrals with variable limits. □

Bernoulli Equations

Bernoulli equations are nonlinear equations having the form

$$y' + p(x)y = q(x)y^n.$$

The transformation of dependent variables $w = y^{1-n}$ turns a Bernoulli equation into a first-order linear equation for w.

Second-Order Equations

Special Equations

Some second-order equations can be immediately reduced to a first-order equation. For example, if the equation has the form

$$G(x, y', y'') = 0,$$

where y is missing, then the substitution $v = y'$ reduces the equation to the first-order equation

$$G(x, v, v') = 0.$$

If the second-order equation does not depend explicitly on the independent variable x, that is, it has the form

$$G(y, y', y'') = 0,$$

then we again define $v = y'$. Then

$$y'' = \frac{d}{dx}y' = \frac{dv}{dx} = \frac{dv}{dy}\frac{dy}{dx} = \frac{dv}{dy}v.$$

So the equation becomes

$$G(y, v, \frac{dv}{dy}v) = 0,$$

which is a first-order equation in $v = v(y)$.

Linear, Constant-Coefficient Equations

The equation

$$ay'' + by' + cy = 0,$$

where a, b, and c are constants, occurs frequently in applications. We recall that the general solution of a linear, second-order, homogeneous equation is a linear combination of two independent solutions. That is, if $y_1(x)$ and $y_2(x)$ are independent solutions, then the general solution is

$$y = c_1y_1(x) + c_2y_2(x),$$

where c_1 and c_2 are arbitrary constants. If we try a solution of the form $y = e^{mx}$, where m is to be determined, then substitution into the equation gives the so-called *characteristic equation*

$$am^2 + bm + c = 0$$

for m. This is a quadratic polynomial that will have two roots, m_1 and m_2. Three possibilities can occur: unequal real roots, equal real roots, and complex roots (which must be complex conjugates).

Case (I). m_1, m_2 real and unequal. In this case two independent solutions are $e^{m_1 x}$ and $e^{m_2 x}$.

Case (II). m_1, m_2 real and equal, i.e., $m_1 = m_2 \equiv m$. In this case two independent solutions are e^{mx} and xe^{mx}.

Case (III). $m_1 = \alpha + i\beta$, $m_2 = \alpha - i\beta$ are complex conjugate roots. In this case two real, independent solutions are $e^{\alpha x} \sin \beta x$ and $e^{\alpha x} \cos \beta x$.

Of particular importance are the two equations $y'' + a^2 y = 0$, which has general solution $y = c_1 \cos ax + c_2 \sin ax$, and $y'' - a^2 y = 0$, which has general solution $y = c_1 e^{-ax} + c_2 e^{ax}$, or equivalently, $y = c_1 \cosh ax + c_2 \sinh ax$. These two equations occur so frequently that it is best to memorize their solutions.

Cauchy–Euler Equations

It is difficult to solve second-order linear equations with variable coefficients. Often, the reader may recall, power series methods are applied. However, there is a special equation that can be solved with simple formulae, namely, a Cauchy–Euler equation of the form

$$ax^2 y'' + bxy' + cy = 0.$$

This equation admits power functions as solutions. Hence, if we try a solution of the form $y = x^m$, where m is to be determined, then we obtain upon substitution the characteristic equation

$$am(m-1) + bm + c = 0.$$

This quadratic equation has two roots, m_1 and m_2. Thus, there are three cases:

Case (I). m_1, m_2 real and unequal. In this case two independent solutions are x^{m_1} and x^{m_2}.

Case (II). m_1, m_2 real and equal, i.e., $m_1 = m_2 \equiv m$. In this case two independent solutions are x^m and $x^m \ln x$.

Case (III). $m_1 = \alpha + i\beta$, $m_2 = \alpha - i\beta$ are complex conjugate roots. In this case two real, independent solutions are $x^\alpha \sin(\beta \ln x)$ and $x^\alpha \cos(\beta \ln x)$.

Particular Solutions

The general solution of the inhomogeneous ODE

$$y'' + p(x)y' + q(x)y = f(x)$$

is

$$y = c_1 y_1(x) + c_2 y_2(x) + y_P(x),$$

where y_1 and y_2 are independent solutions of the homogeneous equation [when $f(x) \equiv 0$], and y_P is any particular solution to the inhomogeneous equation. For constant-coefficient equations a particular solution can sometimes be "guessed" from the form of $f(x)$; the reader may recall that this guessing method is called the method of undetermined coefficients. In any case, however, there is a general formula, called the *variation of parameters* formula, which gives the particular solution in terms of the two linearly independent solutions y_1 and y_2. The formula, which is derived in elementary texts, is given by

$$y_P(x) = \int_a^x \frac{y_1(s)y_2(x) - y_2(s)y_1(x)}{y_1(s)y_2'(s) - y_2(s)y_1'(s)} f(s)ds.$$

There are several introductory texts on differential equations [see, for example, Boyce and DiPrima (1995) or Edwards and Penney (2004)]. Birkhoff and Rota (1978) and Hirsch, Smale, and Devaney (2004) are two more advanced texts.

Exercises

Solve the following differential equations.

1. $y' + 2y = e^{-x}$.
2. $y' = -3y$.
3. $y'' + 8y = 0$.
4. $y' - xy = x^2y^2$.
5. $x^2y'' - 3xy' + 4y = 0$.
6. $y'' + xy'^2 = 0$.
7. $y'' + y' + y = 0$.
8. $yy'' - y'^3 = 0$.
9. $2x^2y'' + 3xy' - y = 0$.
10. $y'' - 3y' - 4y = 2\sin x$.
11. $y'' + 4y = x\sin 2x$.
12. $y' - 2xy = 1$.
13. $y'' + 5y' + 6y = 0$.
14. $y' = \frac{x^2}{1+3y^3}$.
15. $y'' - 6y = 0$.

TABLE OF LAPLACE TRANSFORMS

$u(t)$	$U(s)$
t^n	$n!/s^{n+1}$
e^{at}	$\frac{1}{s-a}$
$\sin at$	$\frac{a}{a^2+s^2}$
$\cos at$	$\frac{s}{a^2+s^2}$
$\sinh at$	$\frac{a}{s^2-a^2}$
$\cosh at$	$\frac{s}{s^2-a^2}$
$u(t)e^{at}$	$U(s-a)$
$H(t-a)u(t-a)$	$e^{-as}U(s)$
$1-\operatorname{erf}\left(\frac{a}{\sqrt{4t}}\right)$	$e^{-a\sqrt{s}}/s$
$\frac{e^{-a^2/(4t)}}{\sqrt{t}}$	$\sqrt{\pi/s}e^{-a\sqrt{s}}$
$\frac{ae^{-a^2/(4t)}}{\sqrt{4t^3}}$	$\sqrt{\pi}e^{-a\sqrt{s}}$
$u^{(n)}(t)$	$s^nU(s)-s^{n-1}u(0)-s^{n-2}u'(0)-\cdots-u^{(n-1)}(0)$
$(u*v)(t)$	$U(s)V(s)$
$tu(t)$	$-U'(s)$
$u(t)/t$	$\int_s^\infty U(r)dr$
$u(at)$	$U(s/a)/a$
$\delta(t-a)$	e^{-as}

References

1. M. Abramowitz and I.A. Stegun, eds., *Handbook of Mathematical Functions*, Dover Publications, New York (1965).
2. R.B. Bird, W.E. Stewart, and E.N. Lightfoot, *Transport Phenomena*, John Wiley and Sons, New York (1960).
3. G. Birkhoff and J.C. Rota, *Ordinary Differential Equations*, John Wiley and Sons, New York (1978).
4. W.E. Boyce and R.C. DiPrima, *Elementary Differential Equations*, 5th ed., John Wiley and Sons, New York (1995).
5. N.F. Britton, Essential Mathematical Biology, Springer-Verlag, New York (2003).
6. H.S. Carslaw and J.C. Jaeger, *Conduction of Heat in Solids*, 2nd ed., Clarendon Press, Oxford (1959).
7. A.J. Chorin and J.E. Marsden, *A Mathematical Introduction to Fluid Mechanics*, 3rd ed., Springer-Verlag, New York (1993).
8. R.V. Churchill, *Complex Variables and Applications*, 2nd ed., McGraw-Hill, New York (1960).
9. R.V. Churchill, *Operational Mathematics*, 3rd ed., McGraw-Hill, New York (1972).
10. L. Edelstein-Keshet, *Mathematical Models in Biology*, McGraw-Hill, New York (1988).
11. C. H. Edwards and D. E. Penney, *Differential Equations*, 3rd ed, Pearson Education, Upper Saddle River, NJ (2004).
12. P. Grindrod, *Patterns and Waves*, Clarendon Press, Oxford (1991).
13. C.W. Groetsch, *Inverse Problems in the Mathematical Sciences*, Vieweg, Braun -schweig/Wiesbaden (1993).
14. R.B. Guenther and J.W. Lee, *Partial Differential Equations of Mathematical Physics and Integral Equations*, Dover Publications, New York (1992).
15. M.W. Hirsch, S. Smale, and R.L. Devaney, *Differential Equations, Dynamical Systems, and an Introduction to Chaos*, 2nd ed, Elsevier (Academic Press), San Diego (2004).

16. M. Kot, *Elements of Mathematical Ecology*, Cambridge University Press, Cambridge (2001).
17. F. John, *Partial Differential Equations*, 4th ed., Springer-Verlag, New York (1982).
18. C.C. Lin and L.A. Segel, *Mathematics Applied to Deterministic Models in the Natural Sciences*, SIAM, Philadelphia (1989).
19. J.D. Logan, *Introduction to Nonlinear Partial Differential Equations*, Wiley-Interscience, New York (1994).
20. J.D. Logan, *Applied Mathematics: Second edition*, Wiley-Interscience, New York (1997).
21. J.D. Logan, *Transport Modeling in Hydrogeochemical Systems*, Springer-Verlag, New York (2001).
22. G. de Marsily, *Quantitative Hydrogeology*, Academic Press, New York (1987).
23. K.W.. Morton and D.F. Mayers, *Numerical Solution to Partial Differential Equations*, Cambridge Univ. Press (1994).
24. J.D. Murray, *Mathematical Biology*, Vol. II, Springer-Verlag, New York (2003).
25. M. Renardy and R.C. Rogers, *An Introduction to Partial Differential Equations*, Springer-Verlag (1993).
26. L.A. Segel, *Mathematics Applied to Continuum Mechanics*, Dover Publications, New York (1987).
27. J. Smoller, *Shock Waves and Reaction-Diffusion Equations*, 2nd ed., Springer-Verlag (1995).
28. W. Strauss, *Introduction to Partial Differential Equations*, John Wiley and Sons, New York (1992).
29. R. Strichartz, *A Guide to Distribution Theory and Fourier Transforms*, CRC Press, Boca Raton (1994).
30. A.N. Tychonov and A.A. Samarskii, *Equations of Mathematical Physics*, Dover Publications, New York (1990).
31. G.B. Whitham, *Linear and Nonlinear Waves*, Wiley-Interscience, New York (1974).

Index

(continued from page ii)

Franklin: Methods of Mathematical Economics.
Frazier: An Introduction to Wavelets Through Linear Algebra
Gamelin: Complex Analysis.
Gordon: Discrete Probability.
Hairer/Wanner: Analysis by Its History. *Readings in Mathematics.*
Halmos: Finite-Dimensional Vector Spaces. Second edition.
Halmos: Naive Set Theory.
Hämmerlin/Hoffmann: Numerical Mathematics. *Readings in Mathematics.*
Harris/Hirst/Mossinghoff: Combinatorics and Graph Theory.
Hartshorne: Geometry: Euclid and Beyond.
Hijab: Introduction to Calculus and Classical Analysis.
Hilton/Holton/Pedersen: Mathematical Reflections: In a Room with Many Mirrors.
Hilton/Holton/Pedersen: Mathematical Vistas: From a Room with Many Windows.
Iooss/Joseph: Elementary Stability and Bifurcation Theory. Second edition.
Irving: Integers, Polynomials, and Rings: A Course in Algebra
Isaac: The Pleasures of Probability. *Readings in Mathematics.*
James: Topological and Uniform Spaces.
Jänich: Linear Algebra.
Jänich: Topology.
Jänich: Vector Analysis.
Kemeny/Snell: Finite Markov Chains.
Kinsey: Topology of Surfaces.
Klambauer: Aspects of Calculus.
Lang: A First Course in Calculus. Fifth edition.
Lang: Calculus of Several Variables. Third edition.
Lang: Introduction to Linear Algebra. Second edition.
Lang: Linear Algebra. Third edition.
Lang: Short Calculus: The Original Edition of "A First Course in Calculus."
Lang: Undergraduate Algebra. Second edition.
Lang: Undergraduate Analysis.
Laubenbacher/Pengelley: Mathematical Expeditions.
Lax/Burstein/Lax: Calculus with Applications and Computing. Volume 1.
LeCuyer: College Mathematics with APL.
Lidl/Pilz: Applied Abstract Algebra. Second edition.
Logan: Applied Partial Differential Equations, Second edition.
Lovász/Pelikán/Vesztergombi: Discrete Mathematics.
Macki-Strauss: Introduction to Optimal Control Theory.
Malitz: Introduction to Mathematical Logic.
Marsden/Weinstein: Calculus I, II, III. Second edition.
Martin: Counting: The Art of Enumerative Combinatorics.
Martin: The Foundations of Geometry and the Non-Euclidean Plane.
Martin: Geometric Constructions.
Martin: Transformation Geometry: An Introduction to Symmetry.
Millman/Parker: Geometry: A Metric Approach with Models. Second edition.
Moschovakis: Notes on Set Theory.
Owen: A First Course in the Mathematical Foundations of Thermodynamics.
Palka: An Introduction to Complex Function Theory.
Pedrick: A First Course in Analysis.
Peressini/Sullivan/Uhl: The Mathematics of Nonlinear Programming.

Prenowitz/Jantosciak: Join Geometries.
Priestley: Calculus: A Liberal Art. Second edition.
Protter/Morrey: A First Course in Real Analysis. Second edition.
Protter/Morrey: Intermediate Calculus. Second edition.
Pugh: Real Mathematical Analysis.
Roman: An Introduction to Coding and Information Theory.
Ross: Elementary Analysis: The Theory of Calculus.
Samuel: Projective Geometry. *Readings in Mathematics.*
Saxe: Beginning Functional Analysis
Scharlau/Opolka: From Fermat to Minkowski.
Schiff: The Laplace Transform: Theory and Applications.
Sethuraman: Rings, Fields, and Vector Spaces: An Approach to Geometric Constructability.
Sigler: Algebra.
Silverman/Tate: Rational Points on Elliptic Curves.
Simmonds: A Brief on Tensor Analysis. Second edition.
Singer: Geometry: Plane and Fancy.
Singer/Thorpe: Lecture Notes on Elementary Topology and Geometry.
Smith: Linear Algebra. Third edition.
Smith: Primer of Modern Analysis. Second edition.
Stanton/White: Constructive Combinatorics.
Stillwell: Elements of Algebra: Geometry, Numbers, Equations.
Stillwell: Elements of Number Theory.
Stillwell: Mathematics and Its History. Second edition.
Stillwell: Numbers and Geometry. *Readings in Mathematics.*
Strayer: Linear Programming and Its Applications.
Toth: Glimpses of Algebra and Geometry. Second Edition. *Readings in Mathematics.*
Troutman: Variational Calculus and Optimal Control. Second edition.
Valenza: Linear Algebra: An Introduction to Abstract Mathematics.
Whyburn/Duda: Dynamic Topology.
Wilson: Much Ado About Calculus.

Breinigsville, PA USA
07 January 2011

252868BV00004B/25/A

9 780387 209531